FROM RECEIVER TO REMOTE CONTROL: THE TV SET

Conceived and Organized by Matthew Geller

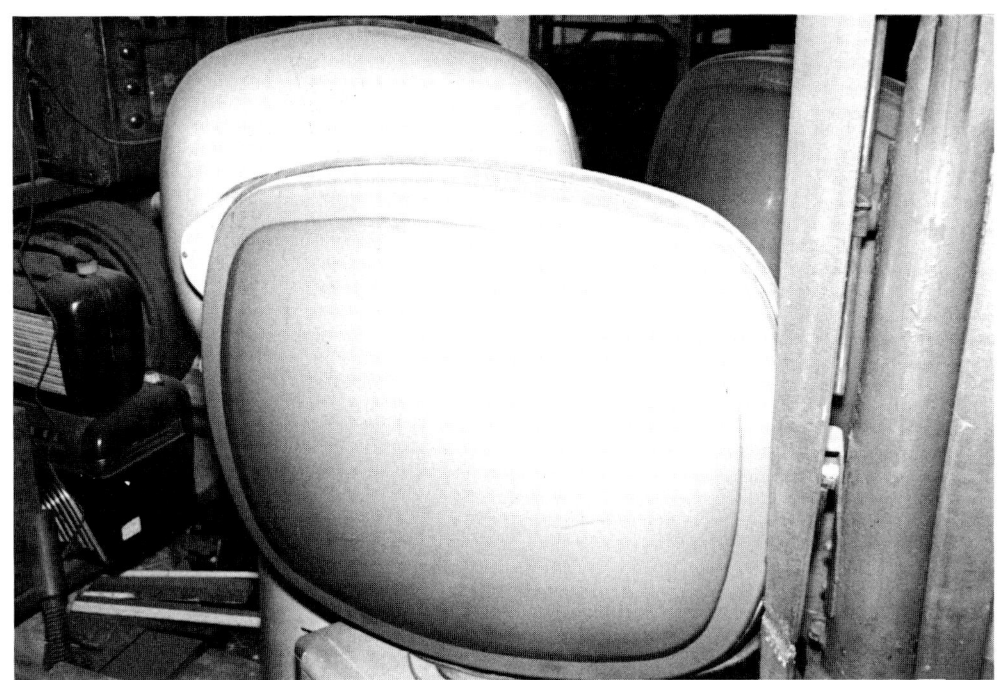

THE NEW MUSEUM OF CONTEMPORARY ART, NEW YORK

From Receiver to Remote Control: The TV Set

The New Museum of Contemporary Art, New York
September 14—November 25, 1990

This book has been edited and produced by Matthew Geller and Reese Williams to accompany the exhibition. It has been printed by Thomson-Shore in Dexter, Michigan.

Library of Congress Catalog Card Number: 90-50390
ISBN 0-915557-70-3

This exhibition has been funded in part by grants from the Design and Museum Programs of the National Endowment for the Arts, the New York State Council on the Arts, the Andy Warhol Foundation for the Visual Arts and the Jerome Foundation. Programs at The New Museum receive operating support from the New York State Council on the Arts, the New York City Department of Cultural Affairs, the office of the Manhattan Borough President, as well as from many individuals, foundations, and corporations.

We gratefully acknowledge the support of the following manufacturers: Bang and Olufsen; Casio, Inc.; Emerson Radio Corporation; Fosgate/Audionics; Mitsubishi Electronic Sales America, Inc.; Monster Cable Products Inc.; Philips Consumer Electonics Company; Pioneer Electronics (USA) Inc.; RCA/Thomson Consumer Electronics, Inc.; Sanyo Fisher USA Corporation; Sharp Electronics Corporation; Sony Corporation of America; Toshiba; Vidikron of America, Inc. and Zenith Electronics Corporation.

We also thank the following lenders to the exhibition: E. Buk Collection, NYC, Arnold Chase, Ned Connors, Jaime Davidovich, Jack Davis, Dan Gustafson, Dave Johnson, Branda Miller and the Museum of Broadcast Communications, Chicago.

Text Credits: "The Domestic Gaze" by Lynn Spigel was originally published as one section of the essay "Installing the Television Set: Popular Discourses on Television and Domestic Space, 1948-1955" in *camera obscura* 16 and is reprinted with the permission of John Hopkins University Press. "The Set in the Sitting Room" by Jane Root was originally published as one section of the book *Open the Box* (Comedia Publishing Group, 1986) and is reprinted with the permission of the author. "How Americans Watch TV: A Nation of Grazers" consists of excerpts from the book of the same title and is reprinted with the permission of C. C. Publishing, an affiliate of ACT III Publishing, New York, NY. "Going to the Movies, Staying at Home" by Kenneth Lawson is © 1989 by The New York Times. The quotation in the first paragraph of William L. Bird's essay on page 63 is reprinted courtesy of the National Broadcasting Company.

Photo and Graphics Credits: *Better Homes and Gardens,* Meredith Corporation, pp. 84, 90, 91; William L. Bird, p. 62; British Film Institute, pp. 32, 33; Courtesy E. Buk Collection, NYC, p. 102; Andrea Callard, p. 123; CCTV Corporation, pp. 92, 93; Peter Collet, front cover; Ned Connors, pp. 104, 129; Matthew Geller, title page, pp. 14, 30, 55; Hitachi Sales Corporation, p. 48; Becky Hunt, p. 38; Basha Kenton for *Audio/Video Interiors,* p. 56; Ehrick V. Long, p. 50; Manfred Montwé, p. 112; Thomas McAvoy, *Life Magazine* © Time Warner Inc., p. 34; Motorola, p. 86; Museum of Broadcast Communications, pp 58, 97; RCA/ Thomson, pp. 20, 61; Sanyo Fisher (USA) Corporation, p. 6; Julia Scher and Chita Contreras, pp. 116, 117; Sharp Electronics Corporation, p. 142; The Smithsonian Institution, pp. 4, 10, 18, 21, 35, 49, 57, 60, 64, 66, 71, 72, 74, 76, 77, 78, 79, 80, 81, 94, 130, 132; Courtesy of *Video Review Magazine,* pp. 19, 36, 37, 44, 59, 75, 109, 118, 124, 128, 134; Paul Warchol pp. 126, 127; Bruce Yonemoto, pp. 82, 83; Zenith Electronics Corporation, pp. 73, 103;

The cover image is a high contrast reproduction of a still frame from a videotape made by Peter Collet which was part of a research study focusing on how people behave while they watch TV.

Contents

Foreword

Marcia Tucker

In 1986, The New Museum invited artists to submit proposals that would use our space, time and resources in a non-traditional way. This initiative resulted in a number of experimental proposals which the Museum wished to pursue in some form. This is one of those projects.

Conceived by artist Matthew Geller and developed under the initiative of The New Museum's curator, the late Bill Olander, *From Receiver to Remote Control* is an unusual exhibition in that it does not consist of art objects, but presents the television set—that ubiquitous object in the American home—in its past, present and future manifestations. Exploring the impact that television's physical presence has had on the American home, family, leisure time and community, this exhibition is devoted to a crucial aspect of contemporary culture. Moving outside a strict definition of the "fine arts," the exhibition addresses the symbiotic relationship of so-called "high" art and "popular" culture.

The accompanying catalogue is also non-traditional in that it does not seek to illustrate the exhibition. Rather, it investigates the impact of the television set through a multi-disciplinary approach and from a variety of viewpoints, addressing a broad and heterogeneous audience.

Our thanks, first and foremost to Matthew Geller, who conceived the project and has carried it through to its conclusion despite difficulties and delays; it is his unique vision and expertise which informs the project throughout. Reese Williams, who has co-edited the book with Matthew Geller, has been responsible for a crucial and major component of the project. Alice Yang at The New Museum has provided the groundwork and coordination for the project since the fall of 1988. Her dedication to the exhibition has been instrumental to its realization.

We are grateful to all the essayists and commentators for contributing their insights to the book; to Barbara Osborn, who provided research for the exhibition and conducted explorative interviews for the book; to Judith Barry and Kenneth Saylor for their extraordinary exhibition design; to Leanne Mella for expert technical direction, without which inevitable chaos would have resulted; and to Branda Miller, for her inspired direction of the Homemade TV project.

On our staff, special thanks to France Morin, Senior Curator, for expertly guiding the staff through this project's challenges, as well as to Debra Priestly, Registrar; Ginny Bowen, Preparator; Wayne Rottman, Gallery/AV Coordinator; Patricia Kirshner, Operations Manager, and our dedicated and hardworking crew for mounting a complex and technically difficult exhibition. Thanks also to Toni DeVito, Director of Planning and Development, for her herculean funding efforts; to Susan Cahan, Education Curator, for her contributions towards an unusual audience exchange; to Sara Palmer, Director of Public Affairs, for her energetic press liaison; and to interns Juliana Engberg and Hilary Gilford for their special assistance.

We are fortunate to have had the invaluable assistance of New Museum Trustee Richard Ekstract and his associates Stephen Booth and Douglas Brod at *Video Review* in assembling equipment for the exhibition. This project would also not have been possible without the generous cooperation of numerous consumer electronics manufacturers and other lenders. Major grants from the NEA's Museum and Design Programs and a planning grant from the New York State Council on the Arts have been instrumental in mounting this exhibition; as always, we are deeply grateful for their support. ■

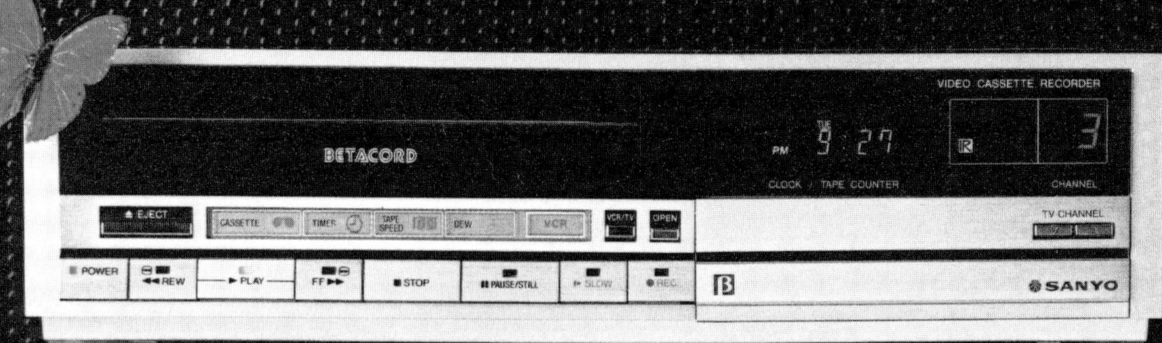
VIDEO CASSETTE RECORDER
BETACORD
CLOCK / TAPE COUNTER
CHANNEL
EJECT
POWER
REW
PLAY
FF
STOP
PAUSE/STILL
SLOW
REC
OPEN
TV CHANNEL
SANYO
MODERN ART TODAY
ARTISTS OF OUR AGE
Design Forum
Design Annual '84
Music and Art Today

SANYO
THE MODERN ART OF ELECTRONICS.

Introduction

Matthew Geller

A riddle: What is it? Every one of us has gained knowledge about it through *personal* experience. It has changed our lives. It is, for most intents and purposes, invisible. It has radically changed relationships among the peoples and countries of the world. And, it's like your first baby—easier to fit into your home if you know something about it. . . .

The exhibition, *From Receiver to Remote Control: The TV Set,* and this book, which has been produced to accompany the exhibition, work together to offer the first comprehensive study of the TV set. There is a vast body of published writing and research on television, but almost all of it focuses on programming, technology, economics, or the history of the television industry. The box itself has largely been overlooked. In one sense, this project takes the TV set's point of view: it looks at itself, it looks at the living room, and it looks at the people watching it.

I spend my time writing, directing or producing videotapes; my instinct is to tell a story. The story of the TV set is more complex, more elusive than it first appears to be. If I were to create a historical chronology and advance a theory about a sequence of cause-and-effect relationships, I am afraid that we would miss the story of the TV set. Lynn Spigel acknowledges the subtle nature of the story when she ends *The Domestic Gaze,* with a quotation from the historian, Carlo Ginzberg: "Reality is opaque; but there are certain points—clues, signs—which allow us to decipher it." Ed Bowes, in his text, *Watching Television* comes to the same point from a slightly different angle: ". . . the working descriptions of things or events are not always neat and well-connected and linear. Modernism recognizes that some complicated things are indeed complicated, and that that's the best way to understand them—to understand them by their details as part of the whole."

My way of telling a story is associative and metaphoric, rather than linear or systematic. Don't look for a central theme. Instead, this book (and exhibition) relies on a set of carefully selected pieces of evidence to communicate the story. Research by distinguished scholars, manufacturers' publicity photographs, personal essays, advertisements for TV sets and paraphernalia, historical photographs, family snapshots, magazine covers, etc. are all offered as evidence. In the end, it is the viewer/reader's personal knowledge and experience that connects the various pieces to create the story.

Back to the riddle for a moment. In just a few decades, the TV set has found its way into almost every home in the country. Ninety-eight percent of the homes in the U.S. have at least one TV set (a higher percentage than homes that have a refrigerator or indoor plumbing). In the average home, the set is on 7.2 hours per day. It's difficult for Americans to think of the TV set as anything more than a standard piece of household equipment. It's omnipresent, yet completely taken for granted. We don't so much watch television as "watch television." We look through the object to the programming it feeds into our homes. The actual set is, for the most part, invisible as we watch it. Part of the work of this project is to identify the visible qualities of the "invisible" by asking questions.

How has the presence of the TV set changed our sleeping patterns and our eating patterns? How has the TV set influenced family and gender relations, and the way we raise our children? How have functional developments in the TV set influenced programming? What does the changing design of the TV set tell us about the illusionary nature of the programming? How did the initial marketing of the set and the viewing habits that followed influence television narrative structure?

For most of us the TV set is the only object in the home that we describe in terms usually reserved for people and pets. A survey revealed our feelings for the set, when it reported that the TV set was the only domestic object that many people said they would not give up for any amount of money. Inmates in a New York State maximum security prison were willing to dramatically modify their behavior when loss of their TV sets was the penalty for not following the rules. Some of us understand the color, hue, contrast and brightness knobs on a TV set, but are confused by our personal affection for the set in a culture which often portrays the set as devilish, alluring and ultimately evil. Can you think of a movie with the TV set as the principal thematic device that does otherwise? Certainly not *Murder by Television* (1933), *Meet Mr. Lucifer* (1953), *Being There* (1979) or *Poltergeist* (1982). The *Channels* publication *How Americans Watch TV: A Nation of Grazers,* excerpted in the book, offers further evidence of this confusion: of the people who responded to a survey by saying they spent "quality time" watching television, twenty percent later admitted that they lied during follow-up telephone calls.

Television has radically changed the "influence = power" equation. What role has the box played in this process? The realm of television is shifting more rapidly than at any time since the TV set first hit the market in the late 1940s. It's a good time to stop and take a look around as we are on the verge of another evolutionary leap. The discussion in the late 1920s concerning which television system was best suited for high definition (later to be known as NTSC), the discourse surrounding the introduction of color television around 1950, and the current high definition (HDTV) debate all bear a striking resemblance. Although the technology is different, the essence of the arguments and the players are basically the same.

We commissioned eleven new essays and interviews for the book and selected excerpts from three previously published works. The essayists were given a few basic parameters and an explanation of the overall shape of the book, but otherwise they were free to address the subject in any way they desired. The result is a wide-ranging collection which crosses the boundaries of communications, sociology, art, business, design and history. The writing is sometimes familiar and reassuring, sometimes unsettling. In Ehrick Long's essay, *A Member of the Family,* Maud Lavin's *TV Design,* Jane Root's *The Set in the Sitting Room* and Ed Bowes's *Watching TV*, for example, we get personal images of the place of the TV in the home and a sense of the low-key affection or attachment that many people have for their TV sets. But in Serafina Bathrick's *Mother as TV Guide,* or Joshua Meyrowitz's *Whatever Happened to Father Knows Best,* the tone is quite different as these two authors explore implications of how the presence of the TV set has affected parent-child relations.

A historical view of the economic issues is highlighted in William Bird's essay *From the Fair to the Family,* while the interview with Les Brown and the excerpt from the *Channels* publication *How Americans Watch TV: A Nation of Grazers* offer some current concerns of "the Industry."

In her essay on the camcorder, Alice Yang introduces the effects of one of the latest technologies to reshape the box—a technology which, in one sense, allows us to place our families in the TV set—the same physical context as the networks. Artists, who were among the first people outside "the Industry" to use portable low-cost video equipment, often skew cultural and technological traditions only to find their work coopted by mass culture a few years later. In his essay, *The Anti-TV,* John Hanhardt introduces some of the earliest works of art (pre-*Porta-Pack*) to decontextualize the TV set.

We live in a time when the older generations remember life without television while the younger ones have never known a time without television. The next generation, or perhaps the following one, may only know the TV set as an object in a museum. They may very well study the TV set in history or sociology class as a way of learning about American culture in the last half of the twentieth century. For example, the family portrait in a typical post-war TV set advertisement isolated Mom from the rest of the family by placing her near the literal edge of the frame. In the 1980s, Mom has been moved to the sofa, but she remains in a subservient position relative to Dad.

The book closes with two essays on the future of the TV set, David Tafler's *I Remember Television. . .* and Margaret Morse's *The End of the Television Receiver*. In 1939, the promise of TV, as it was introduced in advertisements, was that it would offer the equivalent social experience of going to the theater. TV was hailed as a new art form. Just fifty years later, the TV set is beginning to subtly announce its own death. You can see it coming if you watch television—the way the set is being portrayed, immortalized perhaps, in commercials for Toyota MPVs, Kool-Aide, American Express Traveler's Checks, Nickelodeon, 7-Up and others.

Looking back, it seems logical to assume that people would have made a simple transition from the radio to the TV set. The radio would move out of the living room, and the new TV would move in. It wasn't that simple. The confusion and fear that greeted the set's arrival was not so much naivete as it was a premonition of the changes that would follow. Looking ahead, as we move from the TV set into the "virtual space" described in the last two essays, the transition once again appears to be logical. . . .

To bring the story full circle, a few lines of Morse's essay: "The box on display in the living room—the body of television—has been technologically superseded by video projectors and liquid crystal displays. Television no longer requires a three-dimensional body of its own. Though TV sets—be they mahogany or plastic—proliferate numerically into every room in the house and qualitatively beyond, into schoolrooms, courtrooms, legislatures, churches and stadia, the television receiver is already a nostalgic object." ■

FRONT ROW CENTER, U.S.A.

THE lights are dimmed, the voices hushed, the eyes intent as millions of people watch magic through a piece of glass. In town and village, farm and flat, television has made all the world a stage, all living rooms a theatre.

Du Mont is architect of this theatre of the home, the impresario of this new magic. By offering programs to brighten minds, as well as faces, Du Mont has added a new dimension to the entertainment world, a new standard to the art of television.

So, too, has Du Mont added a new dimension to television advertising. Du Mont alone has proved that magic for the viewer need not be fantasy for the sponsor, that truly sensible television means great impact without great cost.

Because it continually seeks to better television, for viewer and sponsor alike, Du Mont will always be a strong force behind the magic behind the glass.

DU MONT TELEVISION NETWORK

515 Madison Avenue, New York 22, N. Y. MU 8-2600
A Division of the Allen B. Du Mont Laboratories, Inc.

The Domestic Gaze

Lynn Spigel

This text is one section of a longer essay, Installing the Television Set: Popular Discourses on Television and Domestic Space, 1948-1955.

While early television programming attempted to fulfill the utopian promise of bringing the world into the home, popular discourses continually deliberated upon the degree to which this new way of seeing could be enjoyed within the domestic context. In some cases, the home was figured as a kind of "ideal theater" where visual pleasures achieved new heights. In fact, the perfect view in television was not only discussed in terms of representational strategies in programming, but also in the context of the home reception environment. In his book, *The Future of Television,* Orrin E. Dunlap, Jr. wrote in 1947 that television was "Utopia for the Audience." Appraising an early NBC drama, he claimed, "The view was perfect—no latecomers to disturb the continuity; no heads or bonnets to dodge. . . In television every seat is in the front row."[1]

The arrangement of the perfect view in the home was constantly discussed in women's home magazines, which advised readers on ways to organize seating and ambient lighting so as to achieve a visually appealing effect for the spectator. In these discussions the television set was figured as a focal point in the home, with all points of vision intersecting at the screen. In 1951, *Good Housekeeping* advised its readers that "television is theatre; and to succeed, theatre requires a comfortably placed audience with a clear view of the stage."[2] Furniture companies like Kroehler "TeleVue" advertised living room ensembles which were completely organized around the new TV center. As this focal point of vision, television was often represented in terms of a spatial mathematic (or geometry) complete with charts indicating optimal formulas for visual pleasure. In 1949, *Better Homes and Gardens* suggested, "To get a good view and avoid fatigue, sit on eye level with screen at no more than 30 degrees off to the side of the screen."[3] Even the TV networks recognized the significance of this new science. CBS in conjunction with Rutgers University studied 102 television homes in order "to determine the distance and angle from which people watch TV under normal conditions."[4]

This scientific management of the gaze in the home, this desire to control and to construct a perfect view, was met with a series of contradictory discourses which expressed multiple anxieties about the ability of the domestic environment to be made into a site of exhibition. The turning of the home into a theater engendered a profound crisis in vision and the positions of pleasure entailed by the organization of the gaze in domestic space. This crisis was registered on a number of levels.

Perhaps the most practical problem which television was shown to have caused was in its status as furniture. Here, television was no longer a focal point of the room; rather it was a technological eyesore, something which threatened to destabilize the unities of interior decor. Women's magazines sought ways to "master" the machine which, at their most extreme, meant the literal camouflage of the set. In 1951, *American Home* suggested that "television needn't change a room" so long as it was made to "retire at your command." Among the suggestions were hinged panels "faced with dummy book backs so that no one would suspect, when they are closed, that this period room lives a double life with TV."[5] In 1953, *House Beautiful* placed a TV into a cocktail table from which it "rises for use or disappears from sight by simply pushing a button. . . ."[6] These attempts to render the television set invisible are especially interesting in the light of critical and popular memory accounts which argue that the television set was a privileged figure of conspicuous consumption and class status for postwar Americans. This attempt to hide the receiver complicates those historical accounts because it suggests that visual pleasure was at odds with the display of wealth in the home.

It wasn't only that the television set was made inconspicuous within the domestic space, it was also made invisible to the outside world. The overwhelming majority of graphics showed the television placed in a spot where it could not be seen through the windows of the room.[7] This was sometimes stated in terms of a solution for lighting and the glare cast over the screen. But there was something more profoundly troubling about being caught in the act of viewing television. The attempt to render television invisible to the outside world was imbricated in a larger obsession with privacy—an obsession which was most typically registered in statements about "problem windows." As discussed earlier, the magazines idealized large picture windows and sliding glass doors for the view of the outside world they provided. At the same time, however, the magazines warned that these windows had to be carefully covered with curtains, venetian blinds, or outdoor shrubbery in order to avoid the "fish bowl" effect. In these terms, the view incorporated in domestic space had to be a one way view.

Television would seem to hold an ideal place here because it was a "window on the world" which could never look back. Yet, the magazines treated the television set as if it were a problem window through which residents in the home could be seen. In 1951, *American Home* juxtaposed suggestions for covering "problem" windows with a tip on "how to hide a TV screen."[8] Even the design of the early television consoles, with their cabinet doors which covered the TV screen, suggested the fear of being seen by television. Perhaps this fear was best expressed in 1949 when the *Saturday Evening Post* told its readers, "Be Good! Television's Watching." The article continued, "Comes now another invasion of your privacy. . .TV's prying eye may well record such personal frailties as the errant husband dining with his secretary. . . ."[9] The fear here was that the television camera might record men and women unawares—and have devastating effects upon their romantic lives.

The theme of surveillance was repeated in a highly self-reflexive episode of the early science fiction anthology, *Tales of Tomorrow*. Entitled "The Window,"[10] the tale begins with a standard sci-fi drama but is soon "interrupted" when the TV camera picks up an alien image, a completely unrelated view of a window through which we see a markedly lower-class and drunken husband, his wife and another man (played by Rod Steiger). After a brief glimpse at this domestic scene, we cut back to the studio where a seemingly confused crew attempts to explain the aberrant image, finally suggesting that it is a picture of a real event occurring simultaneously in the city and possibly "being reflected off an ionized cloud right in the middle of our wavelength, like a mirage." As the episode continues to alternate between the studio and the domestic scene, we learn that the wife and her male friend plan to murder the husband, and we see the lovers' passionate embrace (as well as their violent fantasies). At the end of the episode, after the murder takes place, the wife stares out the window and confesses to her lover that all night she felt as if someone were watching her. As this so well suggests, the new TV eye threatens to turn back on itself, to penetrate the private window and to monitor the eroticized fantasy life of the citizen in his or her home. That this fantasy has attached to it a violent dimension, reminds us of the more sadistic side to television technology as

TV now becomes an instrument of surveillance. Indeed, this fear of surveillance was symptomatic of many statements which expressed profound anxieties about television's control over human vision in the home—especially in terms of its disruptive effects on the relationship between the couple.[11]

Television brought to the home a vision of the world which the human eye itself could never see. We might say that in popular culture there was a general obsession with the perfection of the human vision through technology. This fascination of course predates the period under question, with the development of machines for vision including telescopes, x-rays, photography and cinema. During the postwar period many of these devices were mass produced in the form of children's toys (including microscopes, 3-D glasses, and telescopes) and household gadgets like gas ranges with window-view ovens.

Television, the ultimate expression of this technologically improved view, was variously referred to as a "hypnotic eye," an "all seeing eye," a "mind's eye," and so forth. But there was something troubling about this television eye. A 1954 documentary produced by RCA and aired on NBC suggests the problem. Entitled *The Story of Television*, this program tells the history of television through a discourse on this gaze. A voice-over narration begins the tale in the following way:

> The human eye is a miraculous instrument. Perceptive, sensitive, forever tuned to the pulsating wavelengths of life. Yet the eye cannot see over a hillside or beyond the haze of distance. To extend the range of human eyesight, man developed miraculous and sensitive instruments.

Most prominent among these instruments was the "electronic eye" of television.

In this RCA documentary, the discourse on the gaze was used to promote the purchase and installation of the TV set. However, even in this industry promo, there is something disturbing about the "electronic eye" of television. For here, television inserts itself precisely at the point of a failure in human vision, a failure which is linked to the sexual relations of the couple. A woman frolics on the hillside and we cut to an extreme close-up of a man's face, a close-up which depicts a set of eyes that appear to be searching for the woman. But the couple are never able to see one another because their meeting is blocked by an alternate, and more technologically perfect view. We are shown instead the "electronic eye" of a TV control tower which promises to see better than the eyes of the young lovers. Thus, the authority of human vision, and the power dynamics attached to the romantic exchanges of looks between the couple, is somehow undermined in this technology of vision.

This failure in the authority of human vision was typically related to man's position of power in the domestic space. In 1953, *TV Guide* asked, "Whatever happened to men? Once upon a time (Before TV) a girl thought of her boyfriend or husband as her prince charming. Now having watched the antics of Ozzie Nelson and Chester A. Riley, she thinks of her man as a prime idiot." Several paragraphs later the article relates this figure of the ineffectual male to an inability to control vision, or rather television, in the home. As the article suggests, "men have only a tiny voice in what programs the set is tuned to."[12]

In a 1954 episode of *Fireside Theater*, a filmed anthology drama, this problem is demonstrated in narrative terms. Entitled "The Grass is Greener," the episode revolves around the purchase of a television set, a purchase which the father in the family, Bruce, adamantly opposes. Going against Bruce's wishes, the wife, Irene, makes use of the local retailer's credit plan and has a television set installed in her home.

RCA (CTC-12), 1959. The last RCA color set manufactured without UHF.

When Bruce returns home for the evening, he finds himself oddly displaced by the new center of interest. Upon entering the kitchen door, he hears music and gun shots emanating from the den. Curious about the sound source, he enters the den where he sees Irene and the children watching a TV western. Standing in the den doorway, he is literally off-center in the frame, outside the family group clustered around the TV set. When he attempts to get his family's attention, his status as outsider is further suggested. Bruce's son hushes his father with a dismissive "Shh," after which the family resumes its fascination with the television program. Bruce then motions to Irene who finally—with a look of condescension—exits the room to join her husband in the kitchen where the couple argue over the set's installation. In her attempt to convince Bruce to keep the TV, Irene suggests that the children and even she herself will stray from the family home if he refuses to allow them the pleasure of watching TV. Television thus threatens to undermine the masculine position of power in the home to the extent that the father is disenfranchised from his family whose gaze is fastened onto an alternate, and more seductive, authority.

The crisis was also registered in terms of female positions of pleasure in television. In fact, for the woman, pleasure in viewing television appears to have been "structured absence." These representations almost never show a woman watching television by herself. Typically the woman lounges on a sofa, perhaps reading a book, while the television remains turned off in the room.[13] Two points emerge. First, for women the continuum, visual pleasure—displeasure, was associated with interior decor and not with viewing television. In 1948, *House Beautiful* made this clear when it claimed, "Most men want only an adequate screen. But women alone with the thing in the house all day, have to eye it as a piece of furniture."[14] Second, while these discussions of television were often directed at women, the continuum, visual pleasure—displeasure, was not associated with her gaze at the set, but rather with her status as representation, as something to be looked at by the gaze of another.

On one level here, television was depicted as a threat to the visual appeal of the female body in domestic space. Specifically, there was something displeasurable about the sight of a woman operating the technology of the receiver. In 1955, Sparton Television proclaimed that "the sight of a woman tuning a TV set with the dials near the floor" was "most unattractive." The Sparton TV, with its tuning knob located at the top of the set, promised to maintain the visual appeal of the woman.[15] As this ad indicates, the graphic representation of the female body viewing television had to be carefully controlled; it had to be made appealing to the eye of the observer.

Beyond this specific case, there was a distinct set of aesthetic conventions formed in these years for male and female viewing postures. A 1953 advertisement for CBS-Columbia Television illustrates this well. Three alternative viewing postures are taken up by family members. A little boy stretches out on the floor, a father slumps in his easy chair, and the lower portion of a mother's outstretched body is gracefully lifted in a sleek modern chair with a seat which tilts upward.[16] Here as elsewhere, masculine viewing is characterized by a slovenly body posture. Conversely, feminine viewing posture takes on a certain visual appeal even as the female body passively reclines.

This need to maintain the "to be looked at" status of the woman's body within the home might be better understood in the context of a second problem which television was shown to bring to women—namely, competition for male attention. Magazines, advertisements and television programming often depicted the figure of a man who was so fascinated with the screen image of a woman that his real life mate remained thoroughly neglected by his gaze. Thus, in terms of this exchange of looks, the television set became the "other woman." Even if the screen image was not literally another woman, the man's visual fascination evoked the structural relations of female competition for male attention, a point well illustrated by a cartoon in a 1952 issue of the fashionable men's magazine, *Esquire*, which depicted a newly wed couple in their honeymoon suite. The groom, transfixed by the sight of wrestling on TV, completely ignores his wife.[17] This sexual scenario was also taken up by Kotex, a feminine hygiene company with an obvious stake in female sexuality. The 1949 ad shows a woman who, by using the sanitary napkin, is able to distract her man from his TV baseball game.[18] Perhaps, the ultimate expression of female competition with television came in a 1953 episode of *I Love Lucy* entitled, "Ricky and Fred are TV Fans." Lucy and her best friend, Ethel Mertz, are entirely stranded by their husbands as the men watch the fights on the living room console. In a desperate attempt to attract their husbands' attention, Lucy and Ethel stand in front of the TV set, blocking the men's view of the screen. Ricky and Fred Mertz become so enraged that they begin to make violent gestures, upon which Lucy and Ethel retreat to the kitchen. Having lost their husbands to television, the women decide to go to a drugstore/soda shop. However, once in the drugstore they are unable to get service because the proprietor is likewise entranced by the TV boxing match.

But in what way could this sexual/visual competition appeal to women? A 1952 Motorola ad provides some of the possible answers. The graphic shows a man lounging on a chair and watching a bathing beauty on the TV screen. His wife, dressed in apron, stands in the foreground holding a shovel, and the caption reads, "Let's go, Mr. Dreamer, that television set won't help you shovel the walk." Television's negative effect on household chores was linked to the male's visual fascination in the televised image of another woman. This relationship drawn between the gaze and the household chores only seems to underline TV's negative appeal for women; but another aspect of this ad suggests a less "masochistic" inscription of the female consumer. The large window view and the landscape painting hung over the set suggest the illusion of the outside world and the incorporation of that world into the home. In this sense, the ad suggests that the threat of sexuality/infidelity in the outside world can be contained in the home through its representation on television. Even while the husband neglects his wife and household chores to gaze at the screen woman, the housewife is in control of his sexuality insofar as his visual pleasure is circumscribed by domestic space. The housewife's gaze in the foreground and cited commentary further illustrate this position of control.[19]

This competition for male attention between women and television also bears an interesting relationship to the construction of the female image in domestic comedies. Typically the representation of the female body was de-feminized and/or de-eroticized. The programs usually feature heroines who were either non-threatening matronly types like Molly Goldberg, middle-aged, perfect housewife types like Harriet Nelson, or also zany women like Lucy Ricardo who frequently appeared clown-like, and even grotesque.

Popular media of the postwar years illuminate some of the central tensions expressed by the mass culture at a time when spectator amusements were being transported from the public to the private sphere. At least at the level of representation, the installation of the television set was by no means a simple purchase of a pleasure machine. These popular discourses remind us that television's utopian promise was fraught with doubt. Even more importantly, they begin to reveal the complicated processes through which conventions of viewing television in the home environment and conventions of television's representational styles were formed in the early period.

Magazines, advertisements and television programming helped to establish rules for ways in which to achieve pleasure and to avoid displeasure caused by the new TV object/medium. In so doing they constructed a subject position—or a series of subject positions—for family members in the home equipped with television. Certainly, the ways in which the public took up these positions is another question. How women and men achieved pleasure from and avoided the discomforts of television is, it seems to me, an ongoing and complicated historiographical problem. The popular media examined here allow us to begin to understand the attitudes and assumptions which informed the reception of television in the early period. In addition, they illustrate the aesthetic ideals of middle-class architecture and interior design into which television was placed.

As historian Carlo Ginzburg has argued, "Reality is opaque; but there are certain points—clues, signs—which allow us to decipher it." It is the seemingly inconsequential trace, Ginzburg claims, through which the most significant patterns of past experiences might be sought.[20] These discourses which spoke of the placement of a chair, or the design of a television set in a room, begin to suggest the details of everyday existence into which television inserted itself. They give us a clue into a history of spectators in the home—a history which is only beginning to be written. ■

NOTES

1. Orrin E. Dunlap, Jr., *The Future of Television* (New York: Harper and Brothers, 1947), p. 87.
2."Where Shall We Put the Television Set?" *Good Housekeeping* 133 (August 1951), p. 107.
3. Walter Adams and E.A. Hunferford, Jr., "Television: Buying and Installing It is Fun; These Ideas Will Help," *Better Homes and Gardens* 28 (September 1949), p. 38.
4. Cited in "With an Eye . . . On the Viewer," *Televiser* 7 (April 1950), p. 16.
5. "Now You See It. . . Now You Don't," *American Home* 46 (September 1951), p. 49.
6. *House Beautiful* 95 (December 1953), p. 145.
7. See, for example, *House Beautiful* 91 (October 1949), p. 167; *Better Homes and Gardens* 30 (March 1952), p. 68; *Better Homes and Gardens* 31 (December 1953), p. 71.
8. *American Home* 45 (January 1951), p. 89.
9. Robert M. Yoder, "Be Good! Television's Watching," *Saturday Evening Post* 221 (May 14, 1949), p. 29.
10. Circa 1951-1953.
11. We might also imagine that television's previous use as a surveillance medium in World War II and the early plans to monitor factory workers with television sets, helped to create this fear of being seen by TV.
12. Bob Taylor, "What is TV Doing to MEN," *TV GUIDE* 1 (June 26-July 2, 1953), p. 15.
13. See, for example, *Better Homes and Gardens* 33 (September 1955), p. 59; *Better Homes and Gardens* 31 (April 1953), p. 263; *Popular Science* 164 (February 1954), p. 211; *Ladies Home Journal* (May 1953), p. 11.
14. W.W. Ward, "Is It Time to Buy Television?" *House Beautiful* 90 (October 1948), p. 172.
15. *House Beautiful* 97 (May 1955), p. 131.
16. *Better Homes and Gardens* 31 (October 1953), p. 151.
17. *Esquire* 38 (July 1952), p. 87.
18. *Ladies Home Journal* 66 (May 1949), p. 30.
19. *Better Homes and Gardens* 30 (February 1952), p. 154.
20. Carlo Ginzburg, "Morelli, Freud and Sherlock Holmes: Clues and Scientific Method," *History Workshop* 9 (Spring 1980), p. 27.

"The Chatham" DuMont (RA-103), 1947.

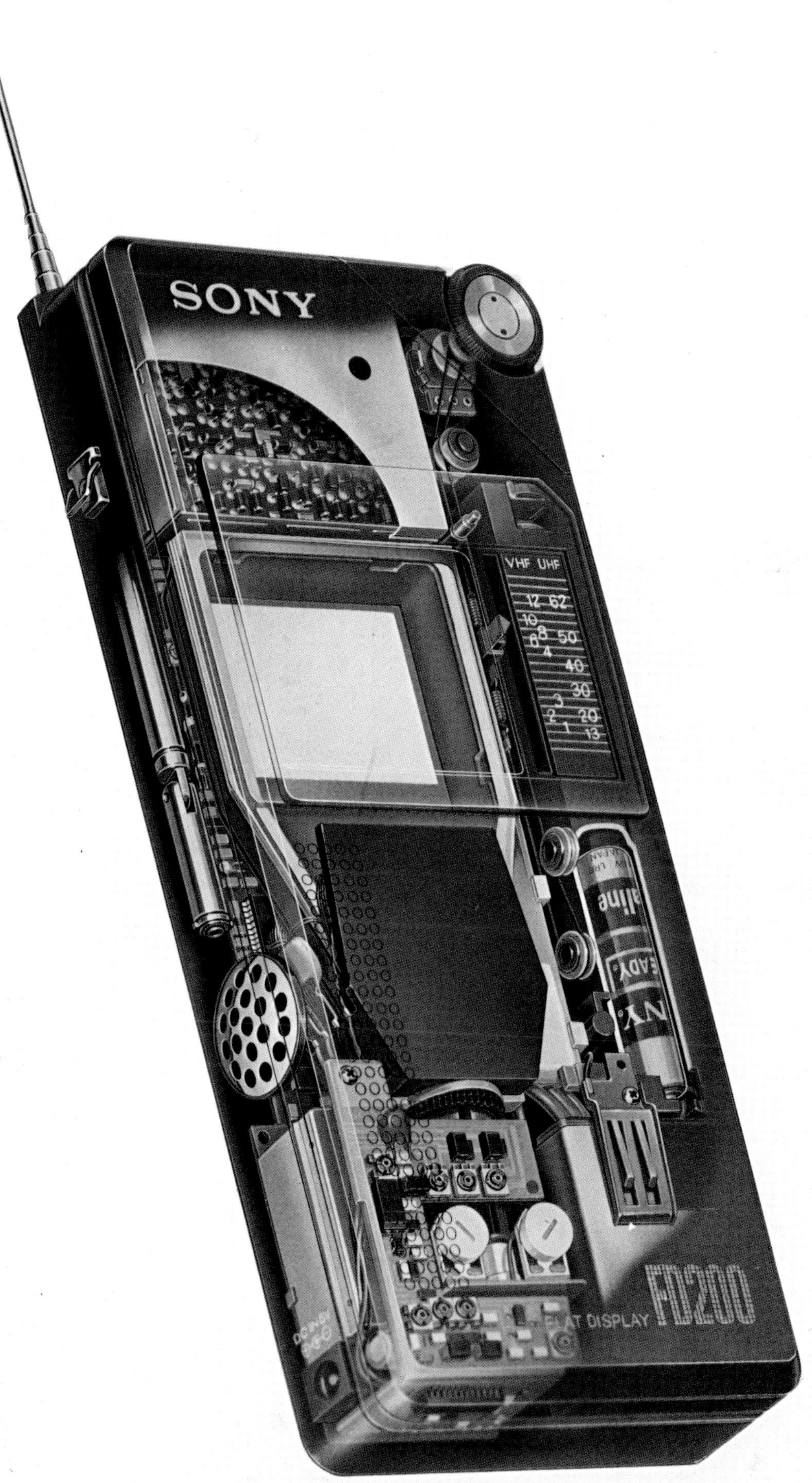
SONY
VHF UHF
12 62
10
8
6 50
4
40
30
3
2 20
1 13
FLAT DISPLAY
FD200

RCA
Solid State
Power
44
DEWEY DEFEATS TRUMA

RCA (621), 1946. Available in Mahogany or Blonde; manufactured for only a few months before being replaced by the 630TS.

Facing page: (top) The 20-pound, 9-inch black-and-white 1976 RCA can be moved indoors or outdoors with its built-in battery, a by-product of space age technology. (bottom) The 85-pound 9-inch 1946 RCA (630TS) was the company's first mass-produced TV set.

No Other Sleeper Can Match

Nitey Nite®

Clever mothers can trim Nitey Nites into a Radish and Watermelon — as seen at this party.

Trust the comfort and health of your young dreamers to the new –improved–NITEY NITE. No other sleeper can match its soft, fleecy fabric–sturdy tailoring–gay songbird colors.

Soft... Fleecy... Fabric

The new fabric, knit of pure cotton, is soft as a downy duckling but there's wear in every thread. Smooth seams are nine-thread sewn; elastic back is bar-tacked. Double sweater-cuffs. Double soled bootee feet. Gripper fasteners.

Perry-ized for Size-Fast Fit

Every NITEY NITE is color-fast and it's size-fast because it's PERRY-IZED. Normal washing can never shrink them out of fit. Two-piece style, also three-piece sets: sizes 00-4. One-piece style: sizes 4-8. Pajama, without feet: sizes 4-14.

Nitey Nite Costumes – Write for the booklet telling how to make children's costumes. Enclose 10¢ for mailing. Dept. P, NITEY NITE SLEEPERS, PERRY, N.Y.

Mother as TV Guide

Serafina K. Bathrick

Appearing in *Parents Magazine* in 1952, an ad for Nitey-Nite pajamas depicts two young siblings playing at a child-sized table and chairs. There is a noisemaker for the older brother and the younger child holds a paper butterfly. Marking the event as a home-grown festivity is the fact that both children wear decorated pajamas and hats that identify them as garden produce. The caption reads: "Clever mothers can trim Nitey-Nites into a radish or a watermelon—as seen at this party." Indeed, one pair of red pajamas has been covered with black "seeds" of fabric and the radish child has sprigs of green attached to her pajama feet.

Parents Magazine always encouraged its readers to invent activities for children, and more specifically, to involve them in activities to transform consumer goods into personalized playthings. Like this ad, the editorials and features of that period promote an ideal of home life that is essentially play-full: a time and place for experiment that might help family players to enter the public world as participants. While the dictates of traditional gender roles are an important aspect of a mother's encouragement, she also inculcates a sense of communal and cooperative play at home.

In 1949, a professor of education writes a prophetic article for *Parents Magazine* in which he warns against some recent developments that seem to mitigate against the acceptance of play within the family.

> In America, too, we are product-minded. We are time-conscious. We want to see results. We are impatient of mistakes. . . This makes it hard for us to feel good about play. For play is messing around; play is experimentation; play is a slow sinking in of life's lessons; play is doing over and over. Play is practice for good living, but we can only tolerate practice when it seems highly directed to some clear-cut end, with the sure promise of quick results.[1]

This educator's position seems particularly relevant when we look at an ad for Nitey-Nite pajamas from a 1957 *Parents Magazine*. Three children are shown sitting together in their nightclothes, staring intently at the same source of light. Posed stiffly as if for a formal portrait, their faces and pastel pajamas glow from an invisible television screen. The copy explains this new family scene in which children are sophisticated spectators: "as seen in the best night spots." The contrast between these two images of childhood suggests an important historical shift that takes place in the mid-twentieth century.

For those of us living in the 1990s to recall our own mothers' relationships with television is to remember three clichés: the mother who screamed "turn off the set and go outside to play"; the mother who turned on the set so as to quiet her screaming children; and the mother who struggled to involve her family in the selection and discussion of programming. This essay is grounded in the assumption that there is a history to these responses, and in particular to the last of them, in which the democratic mother acted with the knowledge that something larger was at stake, beyond the immediacy of her own family's television viewing habits and preferences. It is in the light of her potential to resist the forces of industrialization and consumerism as they eroded democratic ideals that we must consider the subjective experience of women in relation to the mid-twentieth century phenomenon of television.

From the early decades of the nineteenth century, women's magazines offered their readers idealized images of domesticity in which guardian angels and sacrificial mothers reigned over the home. This mythic spirit of motherhood was, however, rooted in some specific historical needs. It reflected an urgency to keep women's powers focused on service to others at a time when the industrial workplace offered an uncaring, if not hostile, environment. While clearly a mid-

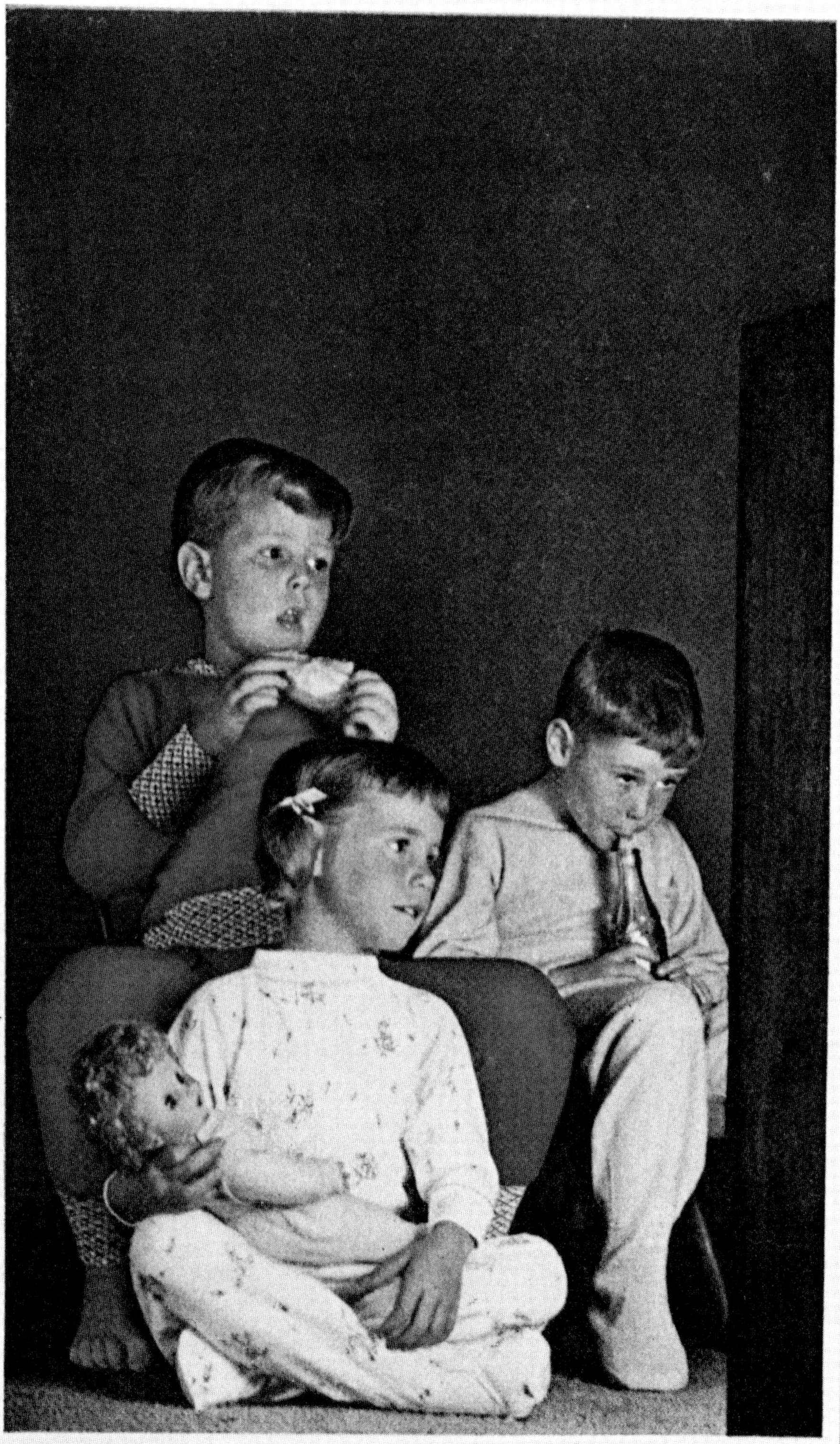

as seen in the best night spots

Wherever you see Nitey Nites, you see happy kids . . . *and* Happy Mamas . . . because these knit sleepers are of fleecy-warm Arctic Cloth that washes and dries fast, needs no ironing, and is Perry-ized to keep its perfect fit. Left to right: Ski Pajamas with bright check trim, 4 to 12; Cupid Print Grow-Sleeper with plastic-soled bootee foot, 0 to 4; Grow-Sleeper with patented bootee foot, sizes 2 to 8. Nitey Nites are priced from 2.25 to 3.98. New, lovable Nitey Nite Doll drinks, wets, has beautiful Saran hair.

they're so snug and elegant in new Arctic Cloth sleepers by

©1957. NITEY NITE SLEEPERS, PERRY, N.Y.

dle class ideal, this image of True Womanhood spoke to all women as they were increasingly exposed to the standards and norms for gendered behavior as defined by the mass media. But in spite of being bound to the concerns of private life, women were asked to play an active role in keeping artisanal values intact, which, along with the participatory values, were integral to the cultural and political promises of a democratic nation. First, the ideal mother was expected to transform her mass produced purchases into personalized goods that met her own family's individual needs. Second, it was within her domain that the ideals of democratic community would survive, and where her children would learn by practice to know themselves and to learn the value of cooperation with others.

Parents Magazine served to strengthen and update for the twentieth century this view of the mother as an historical subject. From the time of its publication in 1926, the editors encouraged readers to define the American family as a democratic community. It is significant that the magazine has its roots in the decade during which the needs of the marketplace mobilized new advertising strategies and an increasing commitment to promote consumerism as a way of life. It also addressed the first generation of American women who could vote and it spoke the language of the efficient workplace to its domestic readers. On the one hand they were expected to take advice from professionals in matters of modern nutrition, psychology and physical fitness, but on the other were encouraged to see themselves as expert consumers when it came to buying goods and inventing personal strategies for their use. In this way, the magazine featured a blend of professional articles and personal testimonials that served to link the modern family with the values of participatory democracy.

This perspective was distinctly different from the narrow ways in which movies and radio programs represented family as a natural, distinctly apolitical community, a place where personal choices were shown to be free from politics or production. These genres of mass entertainment most often made invisible the actual workings of family life, and in so doing sought to cultivate a view of the American mother as the blind and trusting ally of industry and the marketplace. *Parents Magazine,* on the other hand, attributed to women the mindfulness and subjectivity needed by truly historical subjects. In many ways it served to counter the impulse within the mass media to define women as homebodies, separate from community life and untouched by historical change.

But the magazine refused to scapegoat the mass media, presenting instead arguments for the importance of a strong family life which would allow for the coexistence of movies and radio programs with family-generated culture. A regular feature in its monthly issues was a complete list of short reviews and seals of approval for specific Hollywood films deemed suitable for family audiences. The magazine also published articles about the importance of locally based campaigns run by mothers who could demand that the *Parents* seal of approval be accepted by film exhibitors. Nor did this magazine support mindless censorship, as seen when a professional in the field of child development argued against dogmatism and overprotection by adults: "it is quite suitable for parents to loosen their controls over certain areas of the child's experience and to substitute for rigorous censorship, discussions that will make children increasingly intelligent in making choices."[2] By thus placing America's housewives and mothers in powerful positions as the mediators between public and private life, this magazine provided guidance to help families choose mass entertainment which reinforced the same democratic ideals that women themselves upheld.

When television suddenly supplied the mediation between public and private life, *Parents Magazine* did not acknowledge or explore the implications of this harsh substitution. A medium which combined the invasiveness of radio with the visual allure of movies was indeed a powerful challenge to the boundaries of the American household and to the authority of women to bridge the private and public worlds. Never before had the family adapted itself so completely to the rhythms and requirements of consumerism. Never had the sale of goods so directly penetrated the home. At its most extreme, family situation comedies intercut with sales pitches for family-related products threatened to create a closed loop between television and its audience. The authority of housewife and mother to filter and select from the marketplace was thus bypassed by television. This was not just another family purchase that demanded her judgement as to its use and placement within the home. Here was a medium that brought family drama to families and the promise of "life as it happens" to daily life itself. The potential of television to "fit" perfectly into its environment and to flatter home viewers with its own brand of the familiar and the immediate also empowered it to become the new mediation between private and public life. The remaining pages of this essay will explore some of the ways in which this transition was experienced by women. We will see how, without commentary or critique, *Parents* provides a record of how women's authority at home was challenged and usurped by television.

In a poignant article written in 1948 by a housewife named Ella April Codel, *Parents* published its first attempt to focus on how "Television has Changed Our Lives." In one of the few features devoted to the new medium in the period between 1947 and 1957, the author assures her readers that "while family activities go on as usual, television both enriches and entertains."[3] She then reverses her argument by describing several ways in which, following a "family conference" to determine the placement of the set, its presence alters the very rhythms of family life itself.

There is an ironic pleasure communicated as Ella Codel describes her new role as exhibitor/entrepreneur:

> The logical spot in the room was left vacant, awaiting the television installation. I brought down an old table which was not a thing of beauty, but which was the correct size and height on which to place the set. I then made a circular tablecloth with heavy fringe on the bottom to cover the table. In the back of the table I hung ceiling-to-floor draperies of the same fabric to give a simple and unobtrusive effect. The sofa, flanked by lounge chairs, was placed at the opposite end of the room facing the television set.[4]

With the furniture inevitably organized to mirror the mise-en-scené of every living room in every situation comedy, the traditional hearth or even piano are displaced. So too the reorganization of time is acknowledged as an important concession to television programming. The author writes that it is "necessary to be more organized and more efficient in my home." The younger children want to watch a seven o'clock "Small Fry" program so that dinner has to be set ahead. Ella Codel assures her readers that she makes these adjustments gladly and has made TV viewing with the family "a habit, no matter what the program or my chores." A tension between television time and that time which is a part of her household experience or expertise is evident in spite of the author's disclaimers.

> Nightly schedules are pretty well fixed, so I plan accordingly. When a special program is set which I want to see, I usually know about it ahead of time and plan for it. Most of the time it is worth the effort for the satisfaction of witnessing an important event in which I

> could not have otherwise participated.[5]

The Television Age Mother was thus swept into a new role at home. She learned from programs what her family should know about the outside world. As is evident in the above quote, she may even have confused the act of spectating for the act of participation. In spite of her efforts to assure *Parents* readers that a democratic family life was still in order as the group learned to vote on what programs represented a common choice, this author's narrative also recorded a sense of loss. But what makes this substitution of home-made culture by industrial programming so threatening to women and to the democratic values they had upheld is the fact that in their efforts to accomodate the new medium, television culture gained the authority to promote *itself* as an integral feature of family life. In this way, women as homemakers finally became the allies of industry as they facilitated the intrusion of television into their houses. It is possible that just as they had been encouraged by *Parents* to personalize other purchases, so too they brought television sets into their family's lives in the hopes that these little theaters were akin to the puppet shows, costumed plays and readings aloud that had been central to their production of life at home.

Only one other article by a mother describing her experience with television's impact on family life appears in *Parents* between 1947 and 1957. In 1949, Henrietta Battle focuses primarily on the ways in which women should learn to manage meals and family activities around the times of TV programs, and cites some of the rewards for this. Her suggestions for new dimensions in television-time-management at home go far beyond Ella Codel's as she explains her new system:

> My two girls are six and three. At quarter to five they come in from play, literally fly into their bathrobes and slippers, and are ready for supper shortly after five. They eat their meal without dawdling, for they know that 5:30 is "Howdy-Doody Time" (their favorite program). Don't start letting them take their plates with them. They will be too absorbed in the program to eat, and you will lose a valuable way of getting them to clean up their suppers with dispatch. While your children are happily and quietly watching their programs, you can work in the kitchen with peace of mind. — "Television and Your Child" [6]

This new opportunity for precision in the family routine is heightened as the author goes on to explain how, at alternating fifteen minute intervals the three year old and the six year old move like robots between their television programs and their baths and beds, so that while she is putting the younger child down to sleep, "I run the tub for the six year old. She comes up promptly at 6:45 for her bath, and is in bed shortly after 7:00." She notes that any "whining or dawdling" will mean that the child will lose her "television privilege for the next night." The word "dawdle" gains a specific meaning in the age of television, as children are disciplined to program their lives around the set.

These two articles are the only ones to specifically address television's challenge to women's authority at home. What is evident in both of them is that mothers who have been encouraged to view themselves as experts in the area of domestic management are now using that expertise to give over their authority to the demands of the television set. Throughout the 1950s, there is no further discussion of the implications of this substitution or of the subjective feelings of women who must learn to make room for the ways that their families will conform to television programming as they had once conformed to the rhythms of play. That family performance and participation could give way to spectatorship was little anticipated by *Parents*. To follow its approach to television in the 1950s is to see that without a critique of consumer culture, the magazine could not address the new realities of family life colonized and atomized by the values of the industrial marketplace. With little consideration for the ways in which sponsorship based on market research would develop programming that divided the family into specific consumer groups, *Parents* never recognized the challenge television posed to a mother's role as the one who had been encouraged to bring in and personalize goods for her family in the interest of its participatory activities and its community ties. Nor did this family magazine remark on the many ways in which the medium would isolate families at home and thus affect a woman's power to link private and public life through activities and play—the very ways in which the socialization of children had been viewed as an important step in the democratization of individuals. At the same time, the magazine held out for a notion of mothering that asked the impossible of women whose authority at home had been so totally diminished and exploited. That women's time was increasingly divided between work and home was another issue ignored by *Parents,* and in this way alone, we may see that like other mass culture products, it proposed a view of culture that was separate from historical change or from political realities. What is significant here is that it did so at the expense of its own readers.

Parents' inability to approach television as a distinctly different form of mass media is obvious in another article from 1949, and in all of those that followed throughout the first decade of television. A professional writes enthusiastically about the potential for families to find new ways to gather and discuss their interests: the message being that as long as the mother maintains a close watch on her TV spectators, she can continue to monitor and guide their appreciation of democratic values. "Television Comes to Our Children" argues for the many ways that programming is rooted in traditional forms of family-generated entertainment. Puppets, storytellers, and "drawing programs" promise to become popular, and the implication is not that they will replace family-productions, but that they are wholesome because they affirm and re-present that culture. As with earlier articles on how mothers could shape an exhibitor's choices of movies in their local theaters, there is the claim that "programmers would welcome suggestions concerning the age appeal of their present shows, and ideas concerning the likes and dislikes of the various ages."[7] This utopian idea that television was a responsive medium, organized around principles that would permit local community participation, was short-lived. So too the ideal that neighbors would gather to watch together dissolved quickly when every family had their own set.

In the years that follow, *Parents Magazine* devotes a very small percentage of its monthly fare to television related articles, sometimes only one per year and never more than two. These features are written entirely by experts who inform mothers that as long as they are vigilant there will be little danger of delinquency and poor grades posed by the presence of the set. In 1954, a professor of psychology recommends that mothers maintain a "television corner" which can "make viewing productive." So that older children will be inspired to check on the facts that they learn from television, readers are advised to surround the set with encyclopedias and maps. And for younger children who will gain imaginative ideas from their programs, mothers should provide crayons, scissors and "materials needed for participation in activity programs."[8]

There is only one article in the decade between 1948-1958 which refers to commercials on television.

In a throwaway sentence at the end of a list of recommendations to mothers about how to monitor programs and make contact with local stations, two professors suggest comparing notes on the "entertainment value" of a program or talking about "the merits and shortcomings of the story or the acting—or the commercial!"[9] The absence of any discussion on how television programming is linked directly to the promotion and sale of goods surely tells us definitively about the power of this medium to demand that mothers as well as the magazines that have promoted their cause must give in to the new medium. It should also be noted that while *Parents Magazine* continued throughout this period to publish monthly movie reviews and frequent evaluations of all the comic books on the market, there was never such attention paid to television programming. The fact that the set became an integral part of family life almost overnight and that women themselves were shown to have welcomed it, perhaps best explains this phenomenon. Its take-over coincided with a time when women's power to function in every sphere of life was evident and potentially threatening to our gendered notions of social order. Television filled in one of the last remaining gaps that allowed for a relationship between the value of privacy and that of community. It eroded that space and time which had been guarded by women at home in their efforts to resist consumerism and to preserve the power of play and participation within the family. ■

NOTES

1. James L. Hymes, "Why Play is Important," *Parents Magazine* (November, 1949), p. 43.
2. Ernest Osborne, "The Family's Contribution to Democracy," *Parents Magazine* (November, 1937), p. 83.
3. Ella April Codel, "Television has Changed Our Lives," *Parents Magazine* (December, 1948), p. 42.
4. Op. Cit., p. 64.
5. Op. Cit., p. 65.
6. Henrietta Battle, "Television and Your Child," *Parents Magazine* (November, 1949), p. 57.
7. Dorothy McFadden, "Television Comes to Our Children," *Parents Magazine* (January, 1949), p. 76.
8. Robert M. Goldenson, PH.D., "How to Get the Best Out of Television," *Parents Magazine* (November, 1954), p. 128.
9. Paul Witty, PH.D. and Harry Bricker, "Your Child and TV," *Parents Magazine* (December, 1952), p. 76.

Andrew Behar & Sarah Sackner. *TV Dad*, 1988. Color film, 16mm, 29 minutes.

TV Dad

Andrew Behar & Sarah Sackner

Synopsis: *TV Dad,* a thirty minute film completed in 1988, is about a "modern" family—a mother, a father and a son. There is, however, one thing different about this family. The father is dead. In fact, he never even had the chance to meet his wife or son. Knowing that he was dying, the protagonist of this unusual tale depended on artificial insemination and hundreds of videotape cassettes of himself to give his unknown future wife and child the legacy of a husband and a father on TV.

The first scene takes place in a television studio. The protagonist is making a TV advertisement to find a woman who will bear his child. The opening words of the scene are: "By the time you see this, I'll be dead." The scene closes with, "You be Mom, and I'll be TV Dad."

In the next scene, a young woman lounges on the floor in front of the TV, browsing through magazines and eating ice cream. She sees the "TV Dad" advertisement, calls the 800 number and makes an appointment for an interview on the following Tuesday.

The interview consists of a "test" which checks for compatibility with various consumer items—peanut butter, cola, toys, etc. She passes the test, and in quick succession we see: the proposal, the marriage, moving into the new apartment and Mom pregnant.

The apartment has shelves for several hundred videotapes. As the camera scans across the tapes, we see titles like: "Kiddy Show," "Banana Split," "Math" and "Scold" for the child; and "Sex," "Therapy," and "A Quiet Evening at Home" for Mom.

Soon the baby is born, and the scenes that follow offer a sense of what family life is like when "Dad" is always there whenever you need him, "not live, but on videotape."

The scenes that follow (spanning the life of the child from infancy to about age five) include: "Dad" clowning around for the infant, a seduction scene with Mom, teaching the child math, a disciplinary session, a "therapy" session while Mom takes a bath, a fight between Mom and "Dad" and several other scenes.

The film ends "Eight Years Later" with Mom watching a videotape of her son talking to her. ■

Anthony Pelissier, *Meet Mr. Lucifer,* 1953. Black-and-white, sound, 35mm. (above and facing page) Early in the film, a stage actor (Stanley Holloway) performs a play to an almost empty theater—the audience is at home with television. The actor gets drunk and is accidently knocked unconscious during his trap-door entrance from beneath the stage. While unconscious he has a fantasy that he visits Hell and meets the Devil. He learns that television is "an *instrument* of the devil, a mechanical device to make the human race utterly miserable" and concludes that making people aware of this will bring them back to the theater. Many scenes later, after the "Devil's instrument" has been passed among several people, damaging relationships and creating disillusionment, it is thrown in the garbage and a happy audience returns to the theater, but this time to watch a 3-D movie where they can recognize fiction and illusion, unlike viewers at home in their living rooms who are victims of fake togetherness, false community and an ersatz relationship with the performer.

Bush

A NEW 3-BEDROOM HOUSE FOR $25 DOWN

WALTER AND AMY PRICE STAND IN BROADHURST OUTLINE AT SHOWING, SURROUNDED BY ALUMINUM WINDOWS AND OTHER EQUIPMENT THEY WILL GET

A SECOND DU MONT
GIVES TO FATHER
A WELL-EARNED REST
WITH NONE TO BOTHER
Father
Relax
SECOND
DUMONT
Relax
SECOND
DUMONT
Two Telesets
You'll soon agree
Bring even more
Variety.
Children, Mother, Dad
Each one.....
May choose a show
and
All have fun.
DUMONT
Mother
A NEW DU MONT
HELPS MOTHER TO
SEE STYLES AND COOKING
ALL BRAND NEW
Relax
SECOND
DUMONT
DU MONT TELEVISION

Becky Hunt. From the *TV Childhood* series, 1985. Gelatin silver print, 20 x 16". Courtesy of the artist.

Whatever Happened to *Father Knows Best?*

Joshua Meyrowitz interviewed by Barbara Osborn

Joshua Meyrowitz is the author of the award-winning *No Sense of Place: The Impact of Electronic Media on Social Behavior* (Oxford University Press, 1985). He has also written for *Newsweek, Psychology Today, TV Guide, Daedalus* and numerous other popular and scholarly publications. A professor at the University of New Hampshire, he argues that television has blurred the traditional distinctions between public and private spheres. This has had a major impact on the family, specifically affecting conceptions of childhood and adulthood, and of masculinity and femininity.

In No Sense of Place *you say that TV and other electronic media have connected the home to the outside world. What impact has this had on the family?*

The family sphere used to be defined by its isolation from the public realm. There was the public male adult realm of "rational accomplishment" and brutal competition, and the private female and child-rearing sphere of home intuition and emotion. The private realm was supposed to be a place of protection, a place for children to be raised isolated from the nasty realities of adult life. For both better and worse, television and other electronic media tend to break down the difference between those two worlds. The membrane around the family sphere is much more permeable. Children are now exposed to many aspects of adult life, and even homebound women are no longer fully isolated from the public realm. TV takes public events and transforms them into dramas that are played out in the privacy of our living rooms, kitchens, and bedrooms.

Parents used to be the channel through which children learned about the outside world. Parents could decide what to tell their children and when to tell it to them. They could give some books to children and hold back others. Until children can read they don't have access to the information in books. Children tend to learn to read in stages. Books provide a kind of natural screening process, where adults can decide what to tell and not tell children of different reading abilities. So we once had a segregation of adult and child knowledge, and a separation of knowledge into year-by-year slices for children of different ages. Television destroys that whole system. It presents the same information directly to children of all ages, without going through adult filters. TV takes our kids across the globe before parents give them permission to cross the street. Children don't necessarily understand everything that they see on television but they are exposed to many aspects of the adult world that parents might not have decided to tell them about.

So television presents a real challenge to adults. While a parent can read a newspaper without sharing it with children in the same room, television is accessible to everyone in the room. And unlike books, television doesn't allow us to flip through it and see what's coming up. We may think we're giving our children a lesson in science by having them watch the Challenger take off, and then suddenly they learn about death, disaster and adult mistakes. We have no way to protect them from that.

Television also doesn't allow adults to discuss privately what to tell or not tell children. With an adult advice book, adults keep secrets from children and keep their secret-keeping secret. Take that same content and put it on *The Today Show* and you have 800,000 children listening in—learning about the very secrets that adults are suggesting we shouldn't tell our kids. More importantly, kids learn the "secret of secrecy," that adults conspire over what to tell or not tell children. They learn that adults are worried and anxious about being parents.

But isn't it just a matter of programs? What if we only had "innocent" programs on TV, like the family sitcoms of the 1950s?

People often misunderstand this in relation to children. They think if you would only put on programs like *Father Knows Best* we would maintain childhood innocence. They mistakenly believe that because the children characters on those shows were very sheltered and innocent, that such shows would keep their own kids innocent. But when I and my generation watched those programs, we, the child viewers, were not that innocent. We saw the parents behaving one way in front of their kids and another way when they were alone.

There's an episode of *Father Knows Best* that illustrates this beautifully. It begins with the children remarking that their parents never seem to argue. In that same episode Jim decides to teach Margaret to drive, which leads to a number of arguments that they try to keep private from the children. Ultimately, however, the arguments become so intense that they find themselves screaming at each other in the living room. Unbeknownst to them, their three children have descended the staircase (in size place order, of course) and are overhearing the fight. Jim and Margaret turn, see the children, and stop in stunned silence thinking they've been exposed. But the children applaud, saying, "Great show. You were just pretending to argue for us because of what we said, right?" Jim and Margaret look at one another not quite knowing what to do, and then say, "Yes, that's what we were doing." And the kids smile, and Jim and Margaret smile, and the show ends.

Now the children on the show have been fooled, but the children watching the show at home see that adults lie to their children, that adults manipulate their behavior for their children to make it appear that they are perfect. This makes real children much more suspicious of real adults. It makes them wonder what their own parents are hiding from them. It teaches them that deception is part of raising children, and it encourages them to try and uncover such deception. So it's a very exposing view of adulthood.

How has this altered parental authority?

I think adults feel somewhat exposed now and they no longer pretend to know everything in front of their children. This is not to say that adults have absolutely lost their authority. Kids look up to adults and want adults to know a lot of the answers. But through television, kids come to realize that adults do have many, many problems: adults fight, adults kill each other, adults cry, adults lose control, and so on. And adults now know that kids know these things about them. Television supports the desire for adult authority but also makes children aware that it's not always there and that kids may even have to help parents gain authority over them. This paradoxical approach to adult authority is seen most dramatically in the public service messages where kids are shown urging an alcoholic parent to become more responsible. TV empowers children to empower their parents.

This doesn't mean that adults should abdicate their authority over children. Adults are more experienced and more knowledgeable. Adults should try to control and discuss what their kids watch on TV. They should also try and maintain authority in their relationships with children in face to face interaction. But the old support system for unquestioned adult authority has been undermined significantly by television. It's just not possible anymore unless you take the TV out of your house and out of the houses of all your children's friends. The old system of innocent childhood can't be maintained. As a result, we're seeing more adultlike children, and more childlike adults. Kids seem to know much more, and adults now reveal to their children the more childish sides of themselves—such as doubts and anxieties—that they once kept hidden.

What's the psychic price of this blurring of adult and child spheres?

That's a complex question. I can give you a partial, historical answer. With the exception of the last 400 years in Western culture, children have not generally been isolated from adult life. There was no such thing as childhood as we have come to think of it. Even as late as the Middle Ages children dressed like adults, drank in taverns, gambled, went to war, and those few who attended schools often went armed. Children were not viewed as creatures in need of protection from adult behavior, and adults were more childlike. The modern notions of childhood and adulthood developed with the spread of literacy. We may miss some of what we think of as the innocence of childhood, but if we think about which is the norm and which is the aberration, our new way of being is more the norm.

You mentioned that the family sphere was once the woman's realm as well as the child's. Are we seeing changes in male and female roles that parallel the changes in childhood and adulthood?

Yes, very definitely. When television first spread throughout the culture many women were homebound housewives. Women were suddenly exposed to all those parts of the culture that men used to tell them they shouldn't think about, know about, and participate in. There's nothing more frustrating than being exposed constantly to activities that you're told are reserved for someone else. So it's not surprising that women have demanded to participate in all those realms that used to be considered exclusively male, such as business, politics, sports and even war.

The greatest impact is on women, but television also affects men. Television is a very emotional medium; it reveals the personal, emotional side of all public figures and events. We see tears well up in the eyes of a president; we see the yawn, the sweat, the nervous twitch. Television has helped men to become much more aware of their emotions, and aware that emotions can't be completely buried. Men are becoming much more involved as fathers. In short, we're seeing a breakdown of the old segregation of the sexes. So the move away from old gender roles is supported by the shared environment of television, even if the content of TV often portrays sexist role models.

Are the changes in age and sex roles you describe happening among all classes and races?

Things don't end up the same way across all classes and religious and ethnic boundaries. There are other powerful forces there which can't be completely overridden—certainly not in one generation. But we do see a general push in the same direction across groups. Take childbirth, for example. It was once the domain of male doctors; the mothers were drugged unconscious, fathers were excluded, and siblings were kept innocent of the facts of birth, but now women are empowered to make their own birthing decisions, fathers are often present at the delivery, and children are much more likely to be informed and involved. And these changes are happening across all social classes and ethnic groups.

The dissolving of boundaries seems to me an indication of a whole new way of looking at society—a society of permeable or lost boundaries. We actually have a physical restructuring of our world. Of course the change is not total or uniform. Sights and sounds now travel at the speed of light, but we still live in particular places, whether bleak inner-city ghettos or lush suburbs or wherever. It's not surprising that differences remain, but what's remarkable is how quickly and broadly perspectives have changed in the same general direction.

We must realize, however, that while TV encourages a kind of blurring of roles, it doesn't provide for the social mechanisms for allowing change to happen. In fact, the initial short term outcome isn't more harmony or equality; if anything, it's more tension and frustration. TV makes us aware of all the places we can't go, all the people we can't be, all the things we can't possess. TV makes us so aware of the larger world that many of us begin to feel unfairly isolated in some corner of it. And even when "liberation" begins to take place, it's not equally distributed. The women's movement, for example, has primarily been a movement of the upper and middle classes. The lower classes have not benefited very much from it. In fact, the way that many middle class women have been able to enter the male realm is by hiring lower class women to clean their houses and take care of their children.

A lot of people argue that television has cut off the family from the outside world and shut off people from real experience. Statistically more and more families stay at home. How do you respond to this?

Television obviously tips the balance between real and vicarious experience toward the vicarious. And the telephone, stereo, computer and fax machine push us even further in that direction. But life doesn't stop there. When people experience things vicariously there often seems to be fewer reasons not to experience them directly. For instance, we used to think that we shouldn't take kids to funerals, but once they've seen 10,000 people killed on television it no longer makes so much sense not to take them to Grandma's funeral. Similarly, vicarious experiences have helped demystify the male realm for women, encouraging them to enter it.

Of course, mediated experience is not the equivalent of live experience. At the same time I don't think that mediated experiences are always terrible and limiting. As long as you have real live interaction with people, there's nothing wrong with being able to select with whom you have live interactions and with whom you have mediated interactions.

We will continue to interact with and through these new technologies in our lives, but I don't think we'll ever stop interacting with people. We will interact more selectively with people. We will continue to have intimate relationships but with somewhat different structures than we've had in the past. We will no longer be tied to those people who live next to us in terms of relying on them and them on us. Even if you live alone physically now, you're no longer living alone socially; you can still be connected with other people. So we'll continue to be human, but in a different way.

The electronic restructuring of the world changes the meaning of being alone and being together for both better and worse. We are now closer to people who are further away, and further away from people who are close. We share more experiences with people who are thousands of miles away, even as there is a dilution of the intensity of the connection with the people who are in our own houses. A son may walk around the house with a *Walkman* on and a loved one may leave an intimate embrace to answer the telephone.

We're also losing whole systems of behavior. As much as I like the idea of equal access, I'm also aware that we've lost a realm of culture by getting rid of, for instance, exclusively male clubs and by breaking down the separation between the male and female and the child and adult worlds. We've lost a set of subcultures of experience whose variety was interesting. We're losing a sense of local experience and identity, a sense of the limits that used to foster security and intimacy. It's often somewhat bewildering and scary, though, on another level, it's clearly liberating and more democratic. In a world with more and more permeable boundaries, there's a dilution of experience and a growing blandness or sameness on the larger scale. At the same time, there's more choice and variety for each individual.

Television is in many ways a consciousness raising instrument and a secret-exposing machine. It has lifted many of the old veils of secrecy that used to exist between people of different ages, sexes, and levels of authority. Certainly many of the powers that be have tried to coopt, adjust and dodge the effects of that, but they haven't been completely successful. For all the flaws and inequalities that our society still has and the attempts to control what we know, average citizens generally know more about each other and the world than they did before, and they are demanding adjustments in behavior and political realities to match that knowledge.

In many ways we've become hunters and gatherers of the electronic age. As we move forward into the future, we regain an aspect of the earliest form of human organization where people were not tied to specific places. In hunting and gathering societies, the male sphere is not segregated from the female sphere, and the children's sphere isn't separated from the adult sphere. Everybody hangs out together. Women are involved in political decision making and men are involved in child-rearing. Male and female power is relatively equal. Children observe adult behaviors and engage in them in play form.

Does anything about these developments concern you?

Families in our society are under a lot of stress. Part of our culture has changed without the rest of it changing. Television has encouraged the desire for behaviors that are not yet supported by the institutions of our society.

The one sphere of society that has been most resistant to the new permeability of social membranes is the business world, which still functions as if its employees, whether male or female, have a wife at home watching the kids and taking care of the household. So we have a difficult situation now: where both husband and wife have careers, each of which relies on two people, one at work and one at home.

This is not possible. It is not a sane life. In order for our society to continue to function, there have to be adjustments in the expectations of what goes on in the public realm. There have to be more flexible work schedules, and a re-negotiation of the commitment to work. I hope we will see an increasing integration of family life and children in the business world just as we've already seen the home become more business-oriented. ■

General Electric (25PM4854L). 1984 "Performance Plus Series."

The Set in the Sitting Room

Jane Root

This text is an excerpt from Jane Root's book, Open the Box, *published in England in 1986.*

Despite early experiments to transmit programs into cinemas, television is a decidedly housebound medium. We don't have to leave the home to see it, as we do cinema: it sits in there waiting for us. Although anti-television writers (and the understandably antagonistic cinema industry) depict television as an ominous intruder, research suggests that the majority of viewers regard their sets rather affectionately, seeing them, in Peter Collet's[1] words, as "part of the family."

The television manufacturing industry has bolstered this attitude by deliberately downplaying the technological nature of television.[2] Once the early years of uncertain reception and home built receivers were over, they began selling television as "part of the furniture." By 1950, for instancc, *Television Weekly* was able to say, "the accent is on television as part of the home furnishing scheme." Advertisements suggested that a television should be chosen to blend with the design of a sitting room, with other units carefully arranged around it. The set itself was transformed into a homely, domestic device with carefully styled cabinets, highly polished veneer finishes, and sometimes, folding doors. Over the years, the sets have undergone a metamorphosis, from handcrafted walnut cabinets, through Bakelite to the sixties colored plastic murphys, presented as an essential element of any fashionable sitting room. But throughout these changes, the television has stayed a "domestic" object which the advertisements argue should merge with the decor of the room and "say something" about personal taste.

Paradoxically, this tendency has been further underlined by the current matte-black hi-tech designs. Although these sets no longer look ashamed of their electronic innards, they are still promoted as part of a unified home furnishing "look." Ideally, we are told, they should sit alongside a whole series of other similarly cool and industrial objects. But, unlike record, tape and compact disk systems, the promotional material for televisions places comparatively little emphasis on technical developments which might give improved reception. Despite the comparatively poor sound from most sets, no manufacturer has really tried to promote sets offering better sound, or additional speakers. TVs, advertisers' research suggests, are chosen either because of reliable brand names, or for the way they look. Customers aren't really interested in thinking about them as technical appliances.

Once the set is purchased or rented, people will tend to use the set in a way that accords with its status as "friendly object," rather than technical equipment. The Peter Collet tapes suggest that very few people try and watch television in "ideal" conditions, with the lights dimmed and the set carefully tuned for prime performance. Indeed television engineers can be heard complaining bitterly about the poor quality of reception viewers are prepared to accept.

In some houses, the process of domestication started by the manufacturers will be further exaggerated by the careful placing of objects, photographs and family momentos on top of the set. A flavor of the lengths viewers will go to in adorning their television cabinets is given in a 1947 issue of the magazine, *Television Weekly*. In a regular feature called "Across the Counter, Some Jottings by a Television Dealer," the writer comments that a customer once asked, "Will it be all right if I put my aquarium on top of my set?". . . Other curious decorations that I have seen poised on top of television cabinets include flora of all species, from miniature palm-trees to cacti; chiming clocks; "perspex" airplanes and pewter pots, an occasional present from Margate; books; dolls; porcelain animals . . . and, believe it or not, a fair-sized Christmas tree complete with tinsel, colored-balls and crackers. It is clear that television sets have other uses besides the obvious ones!

It has become commonplace to regret the way that television sets have replaced coal fires as focal points of sitting rooms. Television, it is argued, has blocked family discussions and stopped people from talking to each other. Peter Collet believes however, that television is more often an aid for conversation than a barrier to it. "Television is *what* people talk about, while it is on, as well as at work the next day. It buttresses social relationships in the sense that it gives people something to discuss. Often, it provides a kind of focus for people to talk about other things." David Morley[3] concurs: "Television is used as a constant supply of common experience between people who don't necessarily know each other very well. . . At its simplest it is an experience that a lot of people have in common across class barriers, across social divisions of various kinds. The talk may be about television, but very often the point of the conversation is that a relationship is being developed and TV is the easiest form of common experience to refer to as the basis for conversation."

Television runs through family life as the substance of discussions, arguments and recollections. Janet Brown, member of one of the families filmed by Peter Collet, comments: "When me and Marie want to have a mother and daughter discussion we will just turn down the television and sit and chat for a couple of hours. I still know what is happening on television, but when I'm having a heart to heart with Marie my sole attention is on her. Actually, a lot of times the program will actually spark off the discussion. We turn it down so we are watching it *and* having a discussion at the same time."

This does not mean that all families talk about and over the television. Unhappy families are very different. In these, television viewing can be much closer to images of zombified families, with conversation outlawed, except for quarrels over the choice of the program. Television becomes an excuse for not talking, a way of shutting other people out.

This use of television was graphically demonstrated by the flood of distressing letters written to *Woman* magazine. Many men, it seems, use television to punish their wives. The women's misery is almost palpable. One, who self-deprecatingly describes her letter as the "meanderings of an average, bored housewife" writes: "I HATE TV AND WISH THAT IT HAD NEVER BEEN INVENTED. I really miss having family arguments and discussions. My husband is a lecturer at the local polytechnic so he spends most of his time talking to or at his students. So when he comes down he just wants to relax and watch TV and I am stuck with the company of our three-year-old, longing for real conversation. But night after night, on goes that wretched switch. If I complain and he does actually switch *off* then I find I'm too angry to talk! Ironic, but true." The letters suggest that for many men, television provides an excuse for not facing up to the problems in relationships. One letter writer says, "Regarding its effect on the family life, TV just ruins it. That's it in a nutshell. To elaborate, however, it can be a big excuse for not facing up to things."

Some members of the anti-TV lobby, such as Marie Winn[4], see television as the direct cause of these domestic traumas. Undoubtedly, it acts as an irritant. But there is little evidence to suggest that these families would be happy ones were it not for the television in the corner: before the days of television, newspapers were blamed in the same way. Marital disharmony isn't caused by television any more than playground violence. Television is a device which can be called upon to play many different roles in domestic interaction. For some groups it can act as social glue, binding people together. In others, it becomes a potent weapon for wielding against other people, particularly marriage partners.

One of the *Woman* correspondents comments perceptively on this. She begins by saying, "TV made my ex-husband idle, as he would just sit and watch anything. It also helped us to drift apart, as conversation was minimal. If I was to interrupt I was snapped at and told to shut up." Interestingly, her letter ends in a rather different vein: "When a family is close in other ways, then TV isn't a threat, it's just another added pleasure and a nice way of resting and putting one's feet up." Her first marriage finished, she now felt that her experience of television was a complicated reflection of the lack of warmth in the relationship, rather than the direct reason for her unhappiness.

Television can be so vital an element of family interaction that therapists are beginning to use it in discussion sessions. Apparently, conflicts about viewing can be very useful in bringing underlying family tensions to the surface. In particular, male disdain for their wive's choice of program and ways of viewing can be a graphic signal of the deep-rooted inequalities which still fester inside most marriages. Even the remote control can be a revealing indication of family dynamics. According to David Morley, during peak viewing hours the device can be "almost like a medieval symbol of power." In the homes of most of the families he interviewed, it sits "on Daddy's chair and is principally his possession. In some of the Peter Collet tapes you see the man holding the automatic control almost like a mace. It is a very condensed and concrete symbol of authority within the family since it gives the power to change what all the other people in the room are watching."

The battles for domestic power going on in the living room are often mimicked by similar rows taking place on the screen. Television's forte is the minutiae of human relationships, the ups and downs of domestic life. In particular, it is skilled at reflecting the detail of our everyday life back at us. Whereas cinema is good at grandeur and epic scale, television characteristically dwells on the small-scale and intimate. This is true even of programs which claim to be about a majestic sweep of history, such as *Brideshead Revisited* and *The Jewel in the Crown*. Really, when it comes down to it, their stories concern personal relationships and families. The heartland of television is as domesticated as a carefully polished, knick-knack adorned walnut television cabinet. ■

NOTES

1. With the aid of a video camera, Peter Collet conducted a study of "watching people watching television."
2. In the few years since this text was written, manufacturers have begun to give more emphasis to technological advances in sound and picture quality in their advertisements. —*Ed.*
3. David Morley has conducted research on television in England and is the co-editor of *Press, Radio and Television* (Comedia #19).
4. Marie Winn is the author of *The Plug-In Drug* (1977) and *Children Without Childhood* (1981).

There is great happiness in television . . . great happiness in the home where the family is held together by this new common bond – television. And for those who would know the fullest measure of television enjoyment, and see its stirring pageant in thrilling clarity, Du Mont laboratories build television's finest instruments . . . the Du Mont receivers. Everything a television set can be, everything it can offer, is yours in a Du Mont. Console, combinations, table models.

Du Mont built the first commercial home television receiver – Du Mont builds the finest.

DUMONT

First with the finest in Television

THE TARRYTOWN BY DU MONT, *with 17-inch Lifetone* picture.*

* Trade Mark

Ad No. 280B
Look Magazine—October 10, 1950
Colliers (Roto)—October 14, 1950

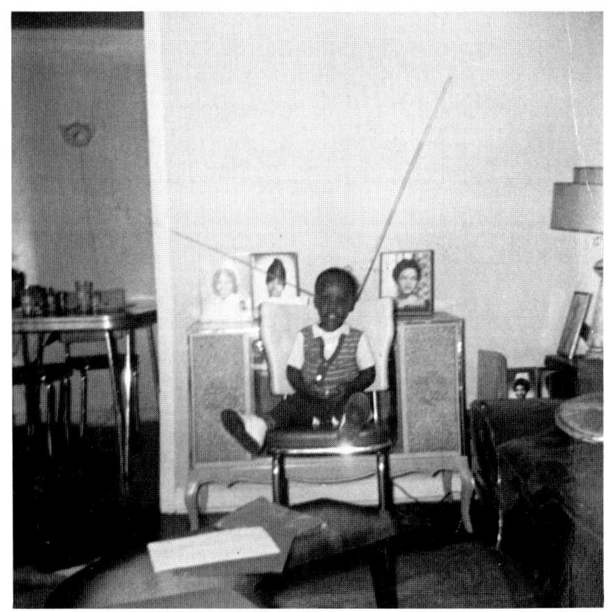

A Member of the Family

Ehrick V. Long

My father, like all fathers I suppose, likes to reminisce about his days as a youth, and my sister and I have endured countless stories about this or that event. However, some of them, such as the ones dealing with the first TV, we have enjoyed all the way up to, say, the ninth or tenth repetition. Apparently his family was the last one on the block to get a TV—not surprising considering the poverty my grandmother faced after losing her first husband to tuberculosis. Although one would figure that other families inhabiting the ghettos of Columbus, Ohio during the mid-'50s would also be in dire economic straits. This set, known as "the TV" in its working days, was a Philco with a large cabinet and a little tiny screen (9 inches). I asked him what they watched in those days and he told me that there were mostly westerns on then: *Tom Mix, Gene Autry, Gunsmoke* etc. They also watched *Howdy Doody,* and Cleveland Browns football games. Reception was not all that good, but it didn't seem to matter. He often said they were so fascinated with TV when it first came out that they would spend hours just watching the test pattern.

My mother, on the other hand, came from a family on the "right side of the tracks." Her parents were comparatively well-off. They had a nice big car and a nice big house in a nice neighborhood filled with nice black professionals. My grandfather, in particular, liked fine things that were beautiful and of high quality. The first TV was no exception. It was an import from a company whose name was too long for me to remember. Its screen (19 inches) occupied one side of a console with a hi-fi and radio on the other. It had so many buttons, control knobs and dials that it looked like it could be programmed to fly to the moon. Maybe that's why my grandfather never let me touch it. Come to think of it, he didn't let my mother touch it too much either. She was allowed to turn it on, but neither she nor my grandmother were permitted to adjust the picture. I remember that I was fascinated with its sliding doors that covered all the electronic stuff, making it look just like the other cabinets in the house. I was equally impressed with the fact that all the controls were labeled in German, although I did not know this until I was old enough to read.

My parents' home was a little different. We had a Zenith that was smaller than the one at both my grandparents' places, and it had an ugly plastic-gray cabinet that didn't match anything else in the house. My mom compensated for it by adding an ugly plastic-green radio to the living room. This helped a little as it diverted some attention away from the TV set. But none of this really mattered much, since we were usually at one of my grandparents' anyway.

As different as my parents' families might seem, there were many similarities in their choice of TVs and the way they used them. First, the TVs were always in the living room. They were too heavy to move about the house, and thus became part of the front-room furniture. Even after the Philco (as it was called in its non-working days) had been replaced by the Curtis-Mathis, it was the place on which photographs, drawings, nice lace cloths, and other things went on display. I know that I considered them to be part of the furniture, mostly because they had a wooden frame that went with the rest of the room's decor—so much so that my cousins and I would climb up and sit on it as we did the tables and the davenport. If my grandmothers hadn't chased us down into the chairs, we would have probably been there all day, as it was our consensus that the TVs were much more comfortable.

It is important to understand, however, that the set was only considered furniture while it was off. When we turned it on, we heard voices. We saw faces. The whole box suddenly became more human. It was only a matter of time before we considered it to be another person. That is not to say that we thought it to be a member of the family—far from it. It was a stranger. Like other strangers, it could be quite entertaining at times, telling us stories of interesting people and places. But there was some distance between us and it. We would never, for example, invite it to a meal in the kitchen, much less allow it into the bedroom. Still, we engaged it in polite conversations, now and then. I, for one, continually asked the set why the cat could never catch the mouse on my favorite Saturday morning cartoon show. It never answered me. Of course, I was a kid, and I was used to this kind of treatment from adults.

But our relationship to the set changed with the advent of the seventies. In 1969, my parents, my sister and I left the rest of the family in Columbus for the last time. Some big move—ninety miles down the road to Cincinnati! Our new house was so beautiful that my mom refused to let either the Zenith or the green radio into the living room. We made a TV-watching space upstairs next to the bedrooms. A month later, we forgot to turn the TV off during a tornado. When the electricity finally came back on, we were already on our way to Atlantic City for a vacation, and by the time we returned home two weeks later all the tubes were burned out.

That was the beginning of the end. We got another TV, a Curtis-Mathis with a built-in stereo. It was a nice contraption with an authentic-looking simulated wood finish. We kept it in the basement for reasons I never quite understood. Perhaps we were just used to having one in the living room. In the basement, it seemed so remote, so far away from where we spent most of our time. To remedy this, we bought a couple of portables, a GE and another Philco, which we moved about from bedroom to bedroom to bathroom to kitchen.

While watching the Reds-Expos game one day, I accidentally spilled a glass of ice water into the Curtis-Mathis. After my dad and I calmly put out the flames, we felt a sense of sadness. Apparently our relationship with the technology had grown more intimate. This set was no longer simply a stranger, it was more like a close acquaintance that died a sudden and violent death.

I wonder sometimes what had happened to make us see the TV in this new light. I'm sure that it was due in part to the portable televisions. After all, it's not as if you could be all that formal in your own bedroom, much less so in the bathroom. But perhaps there was something other than the set's accessibility to more intimate spaces that led to this change in attitude. In the days when TV was an interesting stranger, there was a distance between the images of the people on the screen and us. White middle-class suburbanites such as the Cleavers and the Bradys all seemed like very nice people, but they just were not the kind of folks with whom I could identify. Why aren't there people like me on TV, I wondered. There were, at this time, a few black faces scattered here and there among many shows from *Star Trek* to *The Beverly Hillbillies,* but these roles were hardly the center of attention, nor were they very developed (almost one and one-half dimensional). But with the inclusion of such shows as *The Mod Squad, Room 222,* and *Sanford and Son* most of my family's relationship with TV began to change. Finally, this stranger seemed to be responding to us. It seemed to reflect the knowledge of a friend. Ironically, I occasionally see reruns of some of these shows either on cable TV or at the Museum of Broadcasting, and many of them now strike me as being particularly racist, with many of the stereotyped depictions that people objected to in the days of *Amos 'n' Andy*. But back then it did not matter. We were so starved for representation on the tube that we would have accepted anything.

Nevertheless, the TV was definitely a friend. It was not until some years later (eight, to be exact) that we bought another large piece-of-furniture TV, which despite having a nice wood finish, still does not sit in the living room. Throughout my teenage years, a long procession of small portables marched in and out of our lives with amazing frequency. Hitachi, Sony, Panasonic and all the rest came and went. We mourned their passing and bestowed upon each a proper burial in the middle of the trash heap with full television honors. Usually overheard were the obligatory funeral remarks such as "it had such good color," and "it was only a few years old." My sister once saved up enough money to buy a portable of her own. Unlike our previous TVs, this set could be plugged into the cigarette lighter of the family car. Thus opened a new age of mobility. We took my sister's little TV (I call it that because it still works) on several trips to Wisconsin, Alabama, and the new family home in Ann Arbor, Michigan. In these days, the TV became more like a pet, for we kept it well-fed and full of fresh batteries, but it had to stay in the car overnight.

The TV set eventually did become a member of the family. With the help of some technological breakthroughs, it gained a voice and personality independent from the usual network programming. When my dad bought a VCR and video camera, I no longer had to ask why there was nobody like me on the TV because I was on TV along with the rest of my family. Thus, I can say that this machine even acquired some of our physical traits.

Even more astounding was the inclusion of the *QUBE*-cable box. The *QUBE* system[1] allowed each household to communicate directly with the various cable channels by means of a few buttons. There were game shows and amateur talent contests that you could judge by punching in. My dad also got the pay-per-view attachment complete with lock and key. Turn the key, and the cable people would know that you had the set tuned into a premium channel. It was only then that we became aware that, in a very real way, the TV was watching us while we were watching it. Now, the faces and images that we saw on the screen could (at least theoretically) directly respond to us. This gave credence to everything that we had always felt about the set. It is as though we always expected it to come out of its silence one of these days. Not only that, but with the pay-per-view thing, the cable box, and the *QUBE* attachment (not to mention the VCR, the computer, and the Nintendo), the TV seemed to be growing new appendages. Indeed the set was becoming more human day by day.

My family's relationship with the technology has developed even further over the past few years. You would think that by now, the TV could not really become more intimate in our lives than it already has been, but it has. Not just my family, but my friends as well, are getting into the little Sony *Watchmans* and Casio micro televisions. The set is no longer confined to the home or the car and can be found in elevators, check-out counters or anywhere else the body goes. To look at some of the people I know hold the thing, it would appear that they have grown some strange new appendage of their own.

All in all, I don't know how typical my family is in comparison to all others, but that is our story just the same. From furniture to potential body-part, from stranger to the extension of the self, the development of my family's relationship with the box has always been and will always be dynamic, although I do not care to speculate on how much longer this sense of intimacy could continue to increase. I would hate to think that my grandchildren might be cathode-ray-test-tube babies (*in panisonico* fertilization). Still I must entertain the possibility that this day might come. I do not, for example, think it will be too long from now when people will be able to have sex with TV. With such cable venues as Channel J (sex-oriented programming) here in Manhattan, as well as similar stations in the rest of the country, it already seems to be trying to meet us (at least menfolk, anyhow) halfway in this regard. It's only a matter of time before some company figures out a way to solve some of the inherent interface problems such a union would encounter. ■

NOTES

1. The *QUBE* system was the first commercial application of two-way "interactive" cable TV technology. It took the form of a console attached to the TV set that enabled the home viewer to participate in selected programs by pushing one of five "response" buttons.

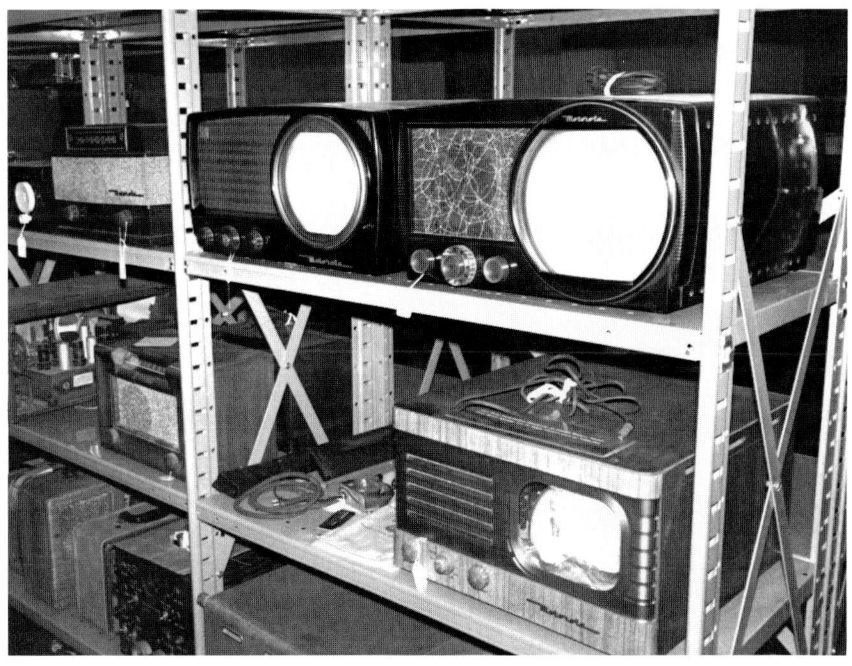

DISPLAY UNTIL APRIL 17

AUDIO/VIDEO INTERIORS

STYLE AND TECHNOLOGY IN HARMONY

VOLUME ONE/NUMBER ONE

Beautiful

NEW 19 inch G-E

Big as you'll ever want!

Here's all you ever dreamed of in a television instrument! Super life-size 19" G-E pictures of breath-taking lifelike quality—result of all the improvements *combined exclusively in G-E Black-Daylite Television.* Now, enjoy pictures so true to life you'll feel they're real! Distinguished cabinetry, veneered in richly grained, genuine mahogany, hand-rubbed to enhance its beauty. New Model 19C105 is truly an enduring investment—in enchantment! **$499.95***

General Electric Company, Electronics Park, Syracuse, N. Y.

**Includes Fed. Tax. Installation and Picture Tube Protection Plan Extra. Prices are slightly higher West and South, subject to change without notice.*

You can put your confidence in—

GENERAL ELECTRIC

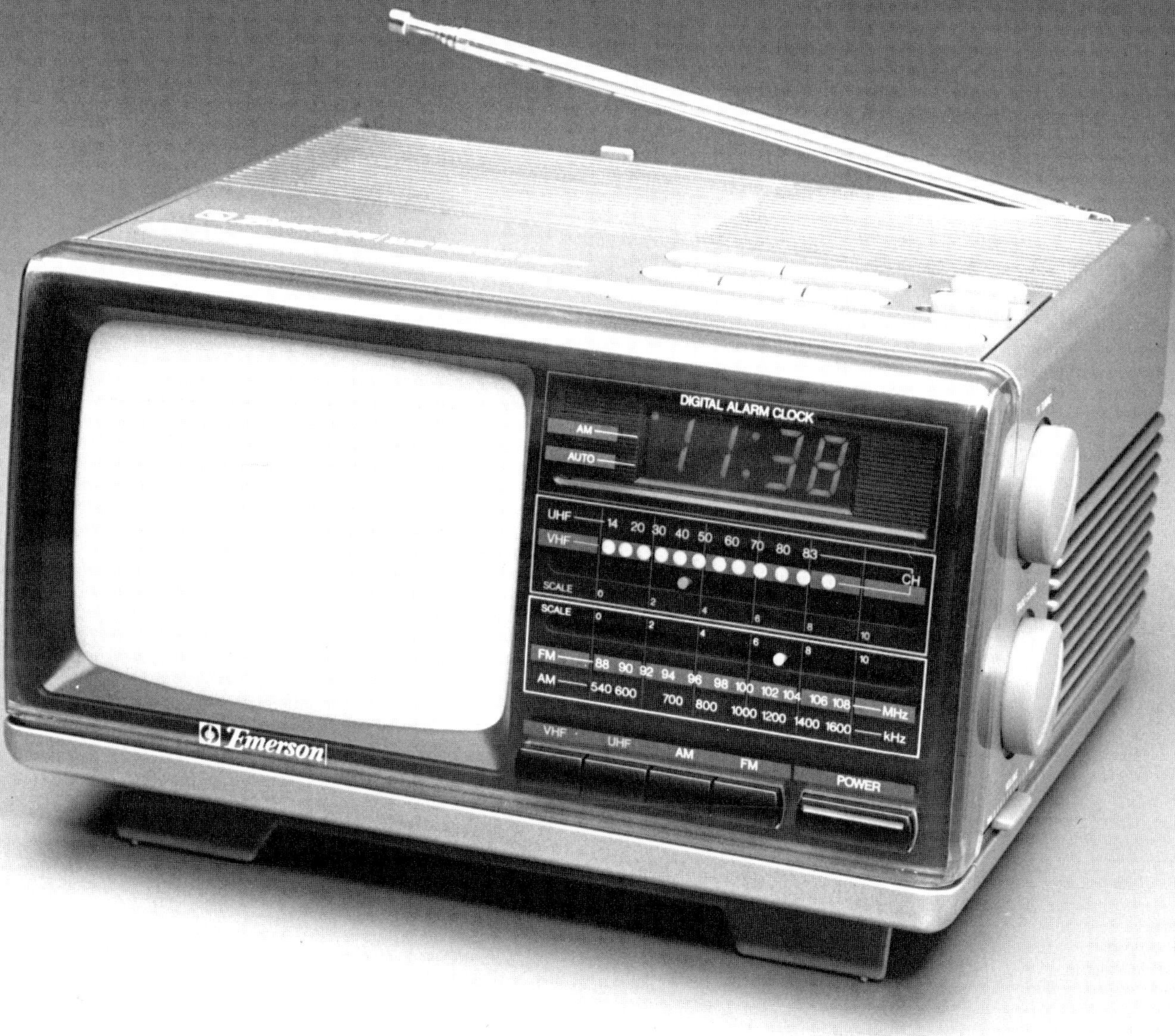
DIGITAL ALARM CLOCK
AM
AUTO
11:38
UHF
14 20 30 40 50 60 70 80 83
VHF
SCALE
CH
FM
88 90 92 94 96 98 100 102 104 106 108
AM
540 600 700 800 1000 1200 1400 1600
MHz
kHz
VHF
UHF
AM
FM
POWER
Emerson

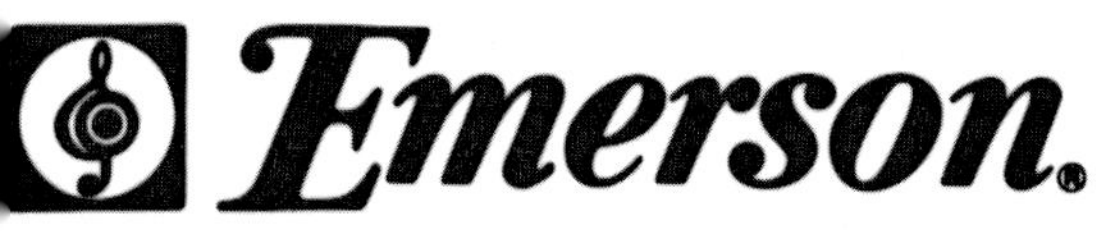

Emerson®

HOW TO LOOK AT TELEVISION—*Continued*

REMOTE CONTROL TV set, designed by Marcel Breuer, custom-built by Philco, is tuned from coffee table. It cost about $900, is on display now at N. Y. Museum of Modern Art.

From *Science Illustrated*, July 1949.

Publicity photograph for the RCA (AM-133), 1970. "The Cheerleader" represents a new portable television design concept. It has a sloping face and back—a design designated "Profile II"—for a new viewing angle. A flip-down rack tilts the screen forward to the conventional upright position as desired.

Ken
Fritz Kuhn Verbatim
MAY 18, 1939
WEEKLY EVERY WEDNESDAY

From the Fair to the Family

William L. Bird

In June 1939, a curious correspondence passed between the executive offices at the Radio Corporation of America. Lenox Lohr, NBC president and the nominal head of marketing of RCA's fledgling television system, outlined for RCA president David Sarnoff a company plan to study the exhibits of the New York World's Fair. "More knowledge," wrote Lohr, "can be gained of an industrial company in an hour's study of its exhibit than in months of questions with agencies and clients, since every product, aim and purpose of the company is visually presented." Warming to an appreciation of its lessons for television, Lohr noted that "the Fair offers an excellent opportunity for the television programming staff to view exhibits which are presented with subtle merchandising and strong entertainment value;. . . since the present exhibits are the best known methods in motivated visual presentations to the ultimate consumer conceived by industry." Lohr particularly liked the short theater show put on in the Westinghouse exhibit entitled "The Battle of the Centuries," that pitted the dishwashing talents of "Mrs. Drudge" against "Mrs. Modern," the operator of a dishwasher.[1]

Though industry's "motivated visual presentations" have since become a television fixture, in 1939 such dramas represented an important change in the way that business talked to the public. At the New York World's Fair and elsewhere during the late 1930s, many of America's largest industrial corporations began dramatizing the personal meaning of their activities and operations in discussions of more and better products, new jobs and preparations for the future. A response to the dislocations of the Depression, the drama of personal meaning stood in contrast to the demonstration of mechanical progress that had animated previous fairs and expositions. It was a propitious time for television, whose technological form neatly fit the stagecraft of corporate social leadership that embraced the objects, advantages and pleasures of home life.

Historians have noted the paradoxical and ironic aspects of the use of the "personal" by an impersonal world to fend off the intrusions that are inevitably its own, and from which there can be only vicarious escape. T.J. Jackson Lears, for example, locates modern advertising's preference for "the personal" in the turn of the century shift from the Protestant ethos of "salvation through self-denial" to "therapeutic self-realization." Roland Marchand, in his study of advertising dating from 1920 to 1940, notes that "modernity" is a condition achieved by a participatory confidence of the "personal," the lubricant of a distended consumer society. Warren Susman, exploring the cultural contradictions of the 1930s, describes the "unanimity of purpose" underlying the concept of the "average American" central to the decade's thinking and planning, whose monuments include opinion polling and Fair architecture that incorporated crowds.[2]

Just as the "personal" smoothed the intrusions of modernity, the dramatization of business' activities and operations in entertaining, ingratiating and above all personally meaningful ways might inoculate them from the regulatory intrusions of the state. The personal touch, however, eluded many in business, despite the encouragement of public relations and advertising specialists who argued that the pro-business publicity of the National Association of Manufacturers, the American Liberty League and the Republican National Committee had become cliched and ineffective in containing the anti-corporate features of the New Deal.[3] The glimmer of a new vocabulary of business leadership appeared with specialists' plea that social leadership become the objective of corporate enterprise in its interpretation to the public. Addressing the

—and JOBS FOR TODAY AND TOMORROW

What Is TELEVISION?

JUST another gadget—another form of entertainment? No. It represents another step forward in man's mastery of time and space. It will enable us, for the first time, to see beyond the horizon. And, in addition, it will create new jobs for today and tomorrow.

New products make new jobs. That's been the history of radio, of the automobile, of electric refrigerators and movie cameras and air conditioning. It's been the history of hundreds of other devices and services that have come from the research laboratories of industry. That's why, in the last 50 years, the number of factory jobs in this country has doubled. And why, in addition, millions of other jobs have been created—selling, servicing, and obtaining raw materials for the new products.

It often takes years of costly, painstaking research to develop a laboratory experiment into a useful product ready for the public to enjoy. This has been the case with television. As long ago as 1930, Dr. E. F. W. Alexanderson and other General Electric engineers demonstrated television to a theatre audience in Schenectady, N. Y. When, after years of labor, television is ready for the public, it will bring to the people of America a new product that will add to their comfort and enjoyment, raise their living standards, and create new employment for today and tomorrow.

G-E research and engineering have saved the public from ten to one hundred dollars for every dollar they have earned for General Electric

NEW YORK—VISIT THE "HOUSE OF MAGIC" AT THE FAIRS—SAN FRANCISCO

National Association of Manufacturers in 1935, for example, Bruce Barton, of the advertising agency Batten, Barton, Durstine & Osborn, whose clients included General Electric and Du Pont, called for a concerted campaign by all business to win popular support for its traditional prerogatives. In language typical of the decade, Barton proposed that business direct its campaign not at Washington but to the people. "Fundamentally," explained Barton, "the people of the United States think they should have a better life, more comfort, more security, more opportunity, more hope. What they are likely to do is to make a choice between industry and politics as to the easiest method of achieving all these benefits." Business, Barton argued, might reclaim its rightful position of social, hence political leadership by investing goods with heightened personal meaning and making them understood as the products of a uniquely productive system of enterprise. Worth quoting for their us-versus-them confidence in the uplifting attributes of consumption, Barton's remarks looked upon the memory of prosperity ushered in with the creation of the automotive and radio industries in the 1920s.

> We say that the automobile business found the poor man chained to his own door-yard, with no horizon but the borders of his own little hamlet, and it has made him the monarch of time and distance. We say that the farm implement industry found man only a little higher than the animals—a valet to horses and chickens and cows; and it leaves him riding like a conqueror over his fields, doing the work of ten men, and yet not too tired for the radio or the movies at night. The electrical industry, the steel industry, the chemical industry, the food industry—dozens of industries—have similar records of acheivement in adding to the comfort, healthfulness, and satisfaction of life. The politician says to the people: "Give us your dollars in taxes, and we will redistribute them." Industry says: "Give us your money in exchange for goods, and we will use it to produce more and better and lower-priced goods." On this issue the competition is joined.[4]

Barton's competitive policy of consumption inspired a stream of company mottoes, each containing the kernel of the idea: Du Pont's "Better Things for Better Living . . . through Chemistry," General Electric's "More Goods for More People at Less Cost" and "More and Better Jobs at Higher Wages," General Motors' "More and Better Things for More People," and General Mills' postwar "New Foods, New Ideas for a Better World."[5] Similar concessions to personal meaning occurred as America's largest industrial corporations entered the entertainment business, and entered to stay as the most expeditious way of asserting their leadership among an uninterested and easily distracted public.

The assertion of corporate leadership held great store for drama in the largest sense of the word, and in the late 1930s propelled to new heights investment in a wondrous popular culture of radio, sponsored films, traveling exhibits, expositions and fairs. The marketing of television dates from this period. Though the television system's economic future was far from certain, on the eve of its introduction coincident with the opening of the New York World's Fair 1939, television had become an important symbol of new jobs, research and preparations for the future based on present science and invention. In language reminiscent of Barton's address to the National Association of Manufacturers in 1935, a General Electric institutional advertisement in February 1939 wondered:

> What Is Television? Just another gadget—another form of entertainment? No. It represents another step forward in man's mastery of time and space. It will enable us, for the first time, to see beyond the horizon. And, in addition, it will create new jobs for today and tomorrow. New Products make new jobs.

ARE YOU READY FOR TELEVISION?

The time is here for America to revise its concepts of its living-rooms, its classrooms, its town halls. The time is here to become familiar with new measurements of human progress... economic, political, scientific.

For full-scale Television is near... a force of unparalleled power. Television will carry new thoughts, new hopes, new products into millions of homes. It will mold men's minds and stir their hearts in a matter of moments. We will watch the truly wonderful tomorrow come into vibrant life before our very eyes.

DuMont will provide you with the finest in Television reception... sight and sound. DuMont quality will be assured by impressive prewar pioneering in Television, by vigorous wartime development, by highly specialized production "know how," by advantageous patents and manufacturing facilities.

Indeed, the world stands on the threshold of an astonishing age... DuMont Television is ready... Are *You?*

DUMONT *Precision Electronics and Television*

ALLEN B. DuMONT LABORATORIES, INC., GENERAL OFFICES AND PLANT, 2 MAIN AVENUE, PASSAIC, N. J.
TELEVISION STUDIOS AND STATION WABD, 515 MADISON AVENUE, NEW YORK 22, NEW YORK

> That's been the history of radio, of the automobile, of electric refrigerators and movie cameras and air conditioning. It's been the history of hundreds of other devices and services that have come from the research laboratories of industry. . . When, after years of labor, television is ready for the public, it will bring to the people of America a new product that will add to their comfort and enjoyment, raise their living standards, and create new employment for today and tomorrow.[6]

Looming above the copy, a cathode ray tube beamed the tipped-in image of workmen approaching a factory. The upper right corner of the advertisement pictured a workman, his sleeves rolled up, assembling a power chasis. It was perhaps the last time that an institutional advertisement featured a television manufacturing scene. Abandoning the point of production for the drama of consumption, subsequent institutional advertisements featured entertainment, sports and abstract impressions of future progress, encased in finely finished cabinetry that aspired to radio's position as the focus of family attention in the home.

Determined to avoid the pitfalls associated with the haphazard development of radio, manufacturers marketed television as a completed system ready to plug in and turn on. Unlike the crystal set days of radio when listeners huddled around coils and wore headsets, television owners would thrill to a "perfected art." DuMont Laboratories, for example, invited viewers to imagine a television experience at home, where "sitting comfortably back in your own living room the lights are dimmed, the viewing screen of your DuMont receiver lights up and a modernistic impression of progress unfolds before your eyes." Manufacturers emphasized their receivers' magical qualities, encumbered by only the intrusion of "a few knobs on the outside of the cabinet. . . necessary to control what would otherwise become a mystifying and complicated device."[7] For many, if not most Americans available to television reception, however, the inordinate cost of receivers (that in the Spring of 1939 ranged from $200 for a "television attachment" to $600 for a "mirror top" television console with radio) deferred the home television experience. Advertisements promoting the sale of receivers typically featured models dressed in formal wear who contemplated small screens picturing similar expressions of class and urbanity. A General Electric publicity still, for example, combining the image of the disposable income required of potential television owners with the image of future progress, pictured two young women in evening dresses enjoying the tipped-in image of the Trylon and Perisphere, the New York World's Fair's thematic centerpiece.

The inauguration of regularly scheduled television program service began in metropolitan New York with the opening of the World's Fair on April 30, 1939. Standing in front of the Federal Building in the Court of Peace, President Franklin D. Roosevelt opened the Fair with a message that was telecast to the roughly 200 television receivers then in metropolitan New York. The number of working television receivers available for public inspection at the inauguration of program service included receivers displayed in the windows and galleries of department stores and radio dealers, in exhibits mounted at the Fair in the Communications Building and in the individual exhibit buildings of General Electric, Westinghouse and RCA.

In demonstrations of television at the Fair, the science and technology of the small screen achieved a certain personal focus that transcended advertisement. At the General Electric, Westinghouse and RCA exhibits, fair-goers could see themselves on closed-circuit television. Some visitors recorded their friends' television appearance in snap shots, and exhibitors issued souvenir cards to their television guests, certify-

John Vassos. *Musicorner*, New York World's Fair, 1940.

ing "________ has been TELEVISED." Exhibits of highly stylized radio-television living rooms invited consideration of the home as the setting, and through television, the location for the dramatization of the world of tomorrow. RCA's "Radio Living Room of Tomorrow," for example, created by John Vassos, the designer of RCA's streamlined television cabinetry, featured a "combination radio, television, record-player and record-recording set, facsimile receiver, and sound motion picture projector. . . built into the furnishings." Exhibited next to it, the "Radio Living Room of Today" decorated in "period furniture" featured "in separate cabinets such as those which are available at present, radio devices for the reception of television, facsimile and sound broadcasting."[8]

The 1940 Fair season saw this vision carried out in the enlargement of television exhibits. The RCA building, enlarged to nearly twice its original size, featured a "Television Suite" of ten separate air-conditioned viewing rooms "furnished as typical American living rooms where television may be seen under circumstances approximating those in the home." Elsewhere at the 1940 Fair, "America At Home," "an exhibit of living in America—around the clock and around the map," featured Vassos' "Musicorner," a room incorporating bleached mahogany modular furniture, indirect lighting, soundproofing, 16mm sound film projector, radio, phonograph, television receiver with pop up screen, and a small library of books and records.[9]

Among the Fair's participants and sympathetic critics, like Gardner Harding, who prepared a preview for *Harper's* in December 1937, the Fair's "vigorous conception of the future. . . may set new bounds to the resources of our daily living." Others, like designer Walter Dorwin Teague, submitted "that better household equipment and better mechanical devices are of no real value unless they are easy first essays in the fundamental redesign of our world: harbingers of a wholesale reorganization of our chaotic scene."[10]

By 1940 that world had taken a decidedly modern cast, and in entertaining and ingratiating ways presented the home as the metaphoric link between industrial civilization and personal meaning in tantalizing proximity to television. Between its television studio and viewing areas, for example, General Electric entertained fairgoers with the "Phantom House," a "fast-moving and amusing demonstration" of "the story of Mr. and Mrs. Tom Morrow, newlyweds, and what happens when Mr. Morrow's mother-in-law comes to visit them." The drama occurred "within a glass house, giving visitors the opportunity to see and listen in on this domestic triangle, and learn how the newlyweds solve the problem raised by the mother-in-law's insistence that she stay with them." The solution involved electric living, that put across "the story of electric appliances for the home with a light touch."[11] In this particular instance television appears not to have entered the picture to speed the solution, but it is hard to recall when, since then, it has not.

Thus television took its place among the wonders of the world, propelled to new heights in the seed time of modern entertainment and ingratiation. Reintroduced after the Second World War, television soon replaced radio as the focus of family attention in the home, and from this unique position assumed the promotional functions of the Fair itself. A symbol of the "world of tomorrow" in 1939, by the early 1950s television had become the predominant medium for the dramatization of that world. ■

NOTES

1. Lenox R. Lohr to David Sarnoff, 1 June 1939, box 79 folder 20: New York World's Fair 1939—1940, central files

correspondence, National Broadcasting Company Collection, State Historical Society of Wisconsin, Madison (hereafter SHSW).

2. T.J. Jackson Lears, "From Salvation to Self-Realization: Advertising and the Therapeutic Roots of the Consumer Culture, 1880-1930," in Richard Wightman Fox and Lears (eds.), *The Culture of Consumption: Critical Essays in American History 1880-1980* (New York: Pantheon, 1983), pp. 1-38; Roland Marchand, *Advertising the American Dream: Making Way for Modernity, 1920-1940* (Berkeley: University of California Press, 1985), pp. 9, 11-12; Warren Susman, "The People's Fair: Cultural Contradictions of A Consumer Society," in Susman, *Culture as History: The Transformation of American Society in the Twentieth Century* (New York: Pantheon, 1984), pp. 211-229.

3. Richard Tedlow, "The National Association of Manufacturers and Public Relations during the New Deal," *Business History Review* 50 (Spring 1976): 25-45; Roland Marchand, "The Fitful Career of Advocacy Advertising: Political Protection, Client Cultivation, and Corporate Morale," *California Management Review* 29 (Winter 1987): 128-156; Frederick Rudolph, "The American Liberty League, 1934-1940," *American Historical Review* 56 (October 1950): 19-33; Burton Bigelow, "Should Business Decentralize Its Counter-Propaganda?" *Public Opinion Quarterly* 2 (April 1938): 321-324; Harold Lord Varney, "Autopsy on the Republican Party," *American Mercury* 40 (January 1937): 1-12.

4. Bruce Barton, "'The Public,' delivered before the Congress of American Industry, in conjunction with the annual convention of the National Association of Manufacturers, December 4, 1935," *Vital Speeches of the Day* 2 (December 16, 1935): 174-177.

5. On the similarity of DuPont's and General Motors' slogans see Paul Garrett to Bruce Barton, 24 January 1944, and Barton to Garrett, 25 January 1944, box 77, folder: General Motors Corp., 1940-1956, client correspondence, Bruce Barton papers, SHSW.

6. Advertisement, General Electric Company, "What Is Television?" *The Forum and Century* 101 (February 1939): overleaf; for a later example in a domestic setting see advertisement, General Electric Company, "'Look, Pop! It's a Homer!'" *The Forum and Century* 102 (September 1939): overleaf.

7. Brochure, DuMont Laboratories, Inc., 1939, box 32 sales and trade literature, Allen B. DuMont Collection, Archives Center, National Museum of American History, Smithsonian Institution (hereafter NMAH/SI); press release, Radio Corporation of America, 20 April 1939, box 549 CWC 142-323A, Clark Collection, NMAH/SI.

8. Press release, Radio Corporation of America, 20 April 1939, box 549 CWC 142-323A, Clark Collection, NMAH/SI; press release, Radio Corporation of America, 11 April 1939, collection of the author.

9. Press release, Radio Corporation of America 15 May 1940, "RCA World's Fair Exhibit," box 79 folder 20: New York World's Fair 1939-1940, central files correspondence, National Broadcasting Company Collection, SHSW; photograph caption, "'Musicorner' section of 'America at Home' exhibition at the New York World's Fair, 1940," Museum of Modern Art.

10. Gardner Harding, "World's Fair 1939: A Preview," *Harper's* 176 (December 1937): 129-137; Walter Dorwin Teague, *Design This Day: The Technique of Order in the Machine Age* (New York: Harcourt, Brace and Company, 1940): 1-2.

11. Press release, General Electric Co., "General Electric at the New York World's Fair in 1940," box 1005 printed material distributed by exhibitors, folder: General Electric, New York World's Fair Corporation Collection, New York Public Library; brochure, "The General Electric Building New York World's Fair 1940," collection of the author.

ADVANCE INFORMATION about post-war shopping

MRS. JONES flicks a switch on her television set and tunes in the Shopping Tele-column of the Air. There she sees and hears the day's best buys, after which she will make up her shopping list and go to market—knowing exactly what she wants.

Far-fetched? Not a bit!

Tomorrow's housewives are going to have an opportunity to see products and packages by television right in their own homes . . . in full color, too! Guided by professional shoppers—yes, and television advertisers—they'll know just what to look for. Shoppers will be better informed and more discriminating than they are today.

That's only one of the many remarkable changes to look for after the war. Because science is making almost incredible progress toward a new way of living.

Stores will change. And products. And packages—for greater eye appeal and product protection.

TELEPHONE
WHITEHALL 3-7226

CABLE ADDRESS
"ANDAREL-NEW YORK"

ANDRÉ DE SAINT-PHALLE & CO.

MEMBERS NEW YORK STOCK EXCHANGE

25 BROAD STREET

NEW YORK 4, N.Y.

ANDRE DE SAINT-PHALLE
CAREL VAN HEUKELOM
JOHN W. KURTH
MEMBER NEW YORK STOCK EXCHANGE
LOUIS LIEBENGUTH
JACQUELINE DE SAINT-PHALLE

CORRESPONDENTS:
ALEXANDRE DE SAINT-PHALLE & CIE.
BANQUIERS
9, RUE BOISSY D'ANGLAS
PARIS, FRANCE

March 29, 1949

Mr. Allen B. DuMont, Pres.
Allen B. DuMont Laboratories
2 Main Avenue
Passaic, New Jersey

Dear Mr. DuMont:

Wherever "growth" industries are discussed, Television is mentioned and, we think, with good reason.

TV receiving set production in 1946 was less than 7,000, in 1947 over 178,000, in 1948 approximately 900,000 and it is estimated that 1949 production will reach the 2,000,000 mark. At the end of January of this year there were 436,000 TV sets in use in the New York area alone and, as you are no doubt aware, Television is now revolutionizing methods of advertising and communication as well as our own lives at home and in public places.

Video is commencing to play a major role in our industrial and military economy and is now available to the viewing public through networks connecting fifteen key cities from the East Coast to the Mississippi. Two years ago there were 20 advertisers sponsoring Video programs and now there are 800. TV time rates have been increased 50% in recent months, a further indication of the increasing recognition of Video as an advertising medium.

While the industry as a whole promises to forge ahead, the problem of selecting the individual stocks most likely to benefit from this growth is not easy to solve. We believe the solution may be found by purchasing shares of an open-end investment trust, specializing in the Video field, which has highly qualified technical consultants and advisers and good management. Such an investment company has been organized and is known as Television Fund, Inc. We think you will be interested to compare the company's holdings as shown in the prospectus of last year with those given in the Report to Stockholders as of January 31, 1949.

Without obligation on your part, we shall be pleased to answer any question you may have pertaining to this type of investment which, in our opinion, offers a sound opportunity to participate with diversification, flexibility and expert supervision in the unlimited potentialities of the Television industry.

Yours very truly,

André de Saint-Phalle & Co

Enc: Prospectus
Stockholders' Report
Inquiry Card

ZENITH
7

DuMont sales conference to introduce the 1953 line of DuMont TVs.

Previous page: Publicity photograph for Zenith "Stratosphere" (L2894U), 1955.

Sony "JumboTRON" at the Crystal Cathedral in Garden Grove, California, 1986.

A television set, golf clubs and a rifle are among the gifts received by Charlie Keller on "his day" at Yankee Stadium, September 25, 1948.

DUMONT

THE WORLD'S LARGEST DIRECT-VIEW SCREEN

Annouucing the Opening of the World's first OUT-DOOR DUMONT TELEVISION STADIUM

Count on Goodman's to "do the unusual." So many people have watched the television broadcasts from our indoor television theatre and from our store windows . . . that we decided to build an outdoor stadium to accommodate the many hundreds who wish to see the shows. And it's ready now . . . right next door to Goodman's . . . featuring the world's largest direct-view screen . . . by DuMont, maker of the world's finest television sets. For you, who wish television in your own home, we suggest you visit Goodman's and see our complete selection of DuMont models.

FREE ADMISSION

Goodman's DuMont Television Stadium will be open afternoons and evenings, except Sundays, to show the exciting baseball games of the Yanks, Giants and Dodgers . . . prize fights . . . football games . . . and other special events. You're cordially invited . . . admission free.

TONIGHT'S EVENT

FROM THE MADISON SQUARE GARDEN

CHARLEY FUSARI

vs.

EDDIE GIOSA

Goodman's Outdoor Television Stadium is right next door to our store. Store hours: Open Monday and Thursday evenings, closed Saturdays during the summer.

Goodman's

FURNITURE APPLIANCES

830 Bergen Avenue • Jersey City

Near Journal Square

Jersey Observer, Friday, July 18, 1947

Starr's Tavern
854 Newark Avenue
Jersey City, N.J.

Installed and serviced by
Goodman's

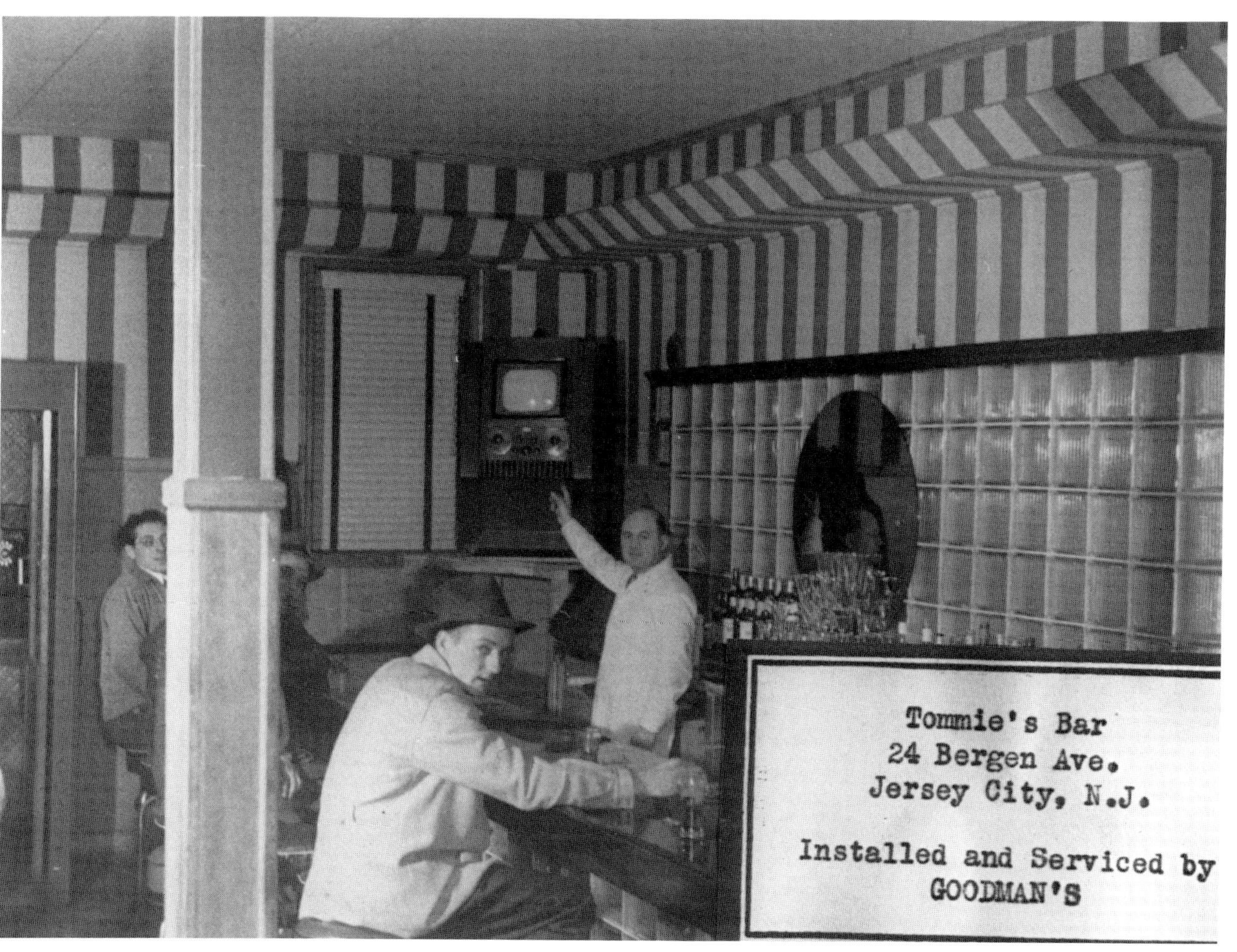
Tommie's Bar
24 Bergen Ave.
Jersey City, N.J.
Installed and Serviced by
GOODMAN'S

OVER 150,000 SATISFIED TELEVIEWERS—*the first year*

Genuine

DELTA-BEAM

Patents Pending

$9.95 LIST

SWIVEL BASE
REVOLVES ON 180° CIRCLE

the World's Most Powerful, All Channel

INDOOR TV ANTENNA

for VHF and UHF Television, and FM Radio

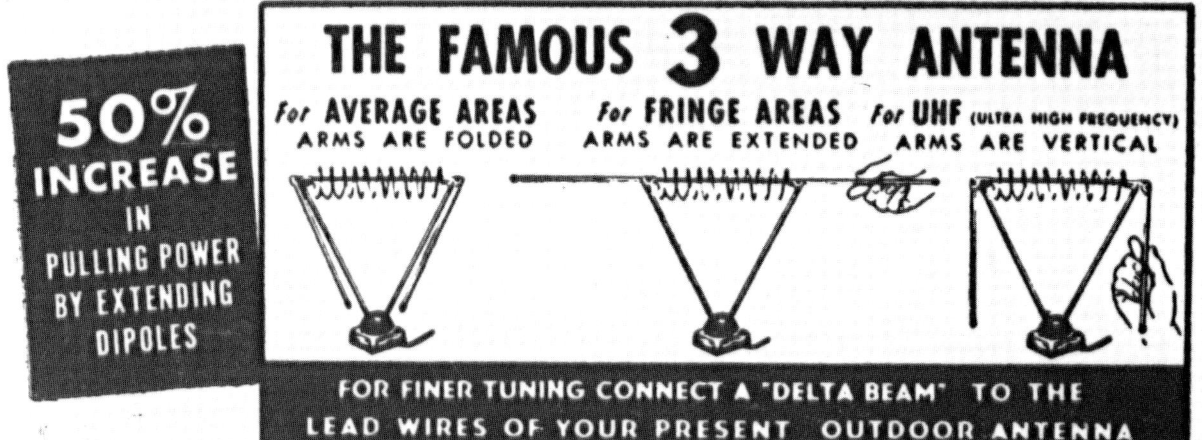

Profit N
Send fo
Full Fa

MFD. BY K-G ELECTRONICS CORP., 2738 N. SHEFFIELD, CHICAGO 14, ILL.

NEW SENSATION IN TV TURNTABLES

BALL-BEARING
Finger-tip turn in full 360° circle

"EASY-WHIRL"

Model No. 14 (at right). Has 24 welds. Holds half-a-ton. Legs 22" high. Individually boxed—Knocked-Down—for regulation P.P. shipping. List 15^{95}

ADJUSTABLE
Strong sleeve adjustment for any TV set up to 32" wide

Black Magic WROUGHT IRON

DISC at base—protects carpets

Smart design STABLE-STURDY

Tempered plastic laminated top. Impervious to burns, stains, etc. Wipes clean.
Model P-104 Mahogany. List
Model P-106 Limed Oak Each **24^{75}**

Extra hard finish lacquered top.
Stain and scratch resistant.
Model L-204 Mahogany List
Model L-206 Limed Oak Each **16^{99}**

24" x 23½". Legs 22" high. Individually boxed, KD.

Table Top in RICH MAHOGANY or LIMED OAK FINISH

Profit Now — Send for Full Information

VINETA MANUFACTURING CO. 2738 N. SHEFFIELD AVE. CHICAGO 14, ILL.

t.v. lamp *n*: a vessel to produce indirect, artificial light prescribed by ophthalmologists during the late 1950's to relieve eye strain due to prolonged television viewing in total darkness-*future archaic:* a surviving vestige [*BIO*. a degenerate or imperfectly designed structure having little or no utility but which in an earlier time performed a useful function] or memorial to an era when a television set was designed as a unit of furniture

Bruce Yonemoto. *TV Lamp Definition*, 1984. Courtesy of the artist.

the best television for the best family in the whole world!

Only RCA Victor brings you the "Magic Monitor" circuit system—

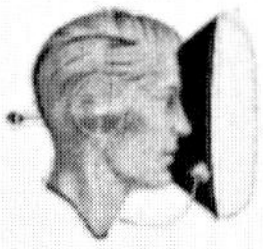

acts like an engineer inside your set

No other TV has so much to offer!

The "Magic Monitor," RCA Victor's exclusive circuit system, screens out static, steps up power, ties clearest picture to best sound —*all automatically*—to bring you the finest TV reception possible.

Fashion-first styling. Authentic period and modern designs so beautifully proportioned and finished that RCA Victor is far and away the favorite for cabinetry.

True, static-free tone from the exclusive "Golden Throat" tone system.

Easy adaptation to UHF because these sets are designed by the same company that pioneered Ultra High Frequency television.

For the ultimate in quality, RCA Victor offers Television *Deluxe*. In these *top* RCA Victor receivers the "Magic Monitor" uses more tubes, more "interference traps" and an extra reserve of power to pull in pictures even in the toughest reception areas, city or country.

***Today*—see RCA Victor Television— and RCA Victor Television Deluxe. It's America's *largest-selling* television!**

Make it your one big family gift—and say Merry Christmas *every* day in the year!

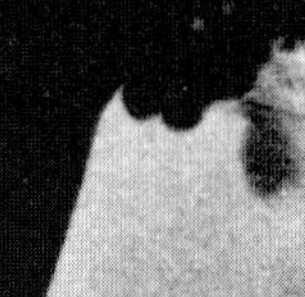

21-inch Sunderland. Finest RCA Victor TV combination. AM/FM radio, "Victrola" 3-speed automatic phonograph and Deluxe 21-inch Television with "Magic Monitor" circuit system—plus *more* "interference traps," *more* reserve power. All in a luxurious cabinet in mahogany finish. Model 21T197DE, $795.00

See the RCA Victor Show, starring Dennis Day over NBC Television, Friday, 8:00 pm., EST.

Give "The Gift That Keeps On Giving" – **RCA VICTOR**

Division of Radio Corporation of America

TV Design

Maud Lavin

When I was growing up in Canton, Ohio, things were so slow that my brothers and I used to entertain ourselves by counting the cars that passed on the road in front of our house. Since we lived on a gravel road well outside of town, only three or four cars passed every day. Another favorite form of entertainment was to collect the bits of gravel thrown up onto the lawn by the cars. My father would pay us a penny each for every gravel chip we gathered. Anyway, you can well understand the impact television must have had against such pastoral forms of entertainment.

When I think about growing up with television, I can remember almost everything about the experience except the appearance of the television set itself. What counted more than the set's design was the way the television redesigned our social space and patterns of interaction. The TV was located in a modern, '50s style playroom complete with a picture window that looked out over the cornfields. The shades were never drawn on the window since we had no neighbors near enough to look in. Often the window was open and while you were watching television you could smell the alfalfa and hear the crickets chirping. I know it sounds like a Norman Rockwell painting or like Dorothy in Kansas, but these sensual experiences were, for me, inextricably bound up with the pleasures of TV viewing.

My parents rarely joined the kids in watching television. My mother would be upstairs reading and my father would only occasionally wander in, restless, to watch a half hour or so of whatever was on. Basically it was just my three younger brothers and me; I have no memories of watching television alone. For us kids watching television was no casual affair. The TV had to be positioned just right. It was a table-top model, and we would draw up our chairs in a precise semicircle, each one as close to the TV as possible while still allowing space for the other three and blocking no one's vision. In true Skinnerian form, my parents devised a complex chart for TV viewing: we were assigned turns as to who had first choice of the TV programs each night and who had first dibs on the favorite chair. (There were other charts in the house about who takes a bath when, who washes his or her hair, does the dishes, cuts the grass—but that's another story). Our interest in television design had to do with sharing and, until the charts were drawn up, competing for the space around the television, a space which was nevertheless ours and not our parents.

Considering how ubiquitous television is in our lives, it is surprising how little attention has been paid to the intimate assimilation of the TV set into our homes in visual and spatial terms. In fact, there are two design histories of television to be constructed, and they have almost nothing to do with one another. One is an official narrative, a chronology of changes in the TV set and the other is unofficial, a collection of personal memories of growing up with TV, telling how the TV set was incorporated into home, family, and leisure time. The two histories, one public, one domestic, cover the same decades, but they rarely intersect. Even in the fifties when the design teams of Motorola and Zenith and GE were so desperately competing for what was fast becoming a saturated market, individual families like mine paid little attention. It was not that consumers lacked interest in product design, but that television, as solid state equipment, hardly ever needed to be replaced. Design, in relation to the television set, meant working the immutable box into our lives.

The field of design history is relatively underdeveloped as a field of study. In the U.S., design writings tend to either heroicize individual designers or focus on technological changes. In fact, most design

Motorola TV CABINET STYLE

Shows how to MATCH or MIX TV Cabinet Styles with

HOME FURNITURE STYLE

PERIOD FORMAL

Georgian | English Regency | Victorian | Queen Anne | Chippendale | French Empire | Duncan Phyfe

PERIOD INFORMAL

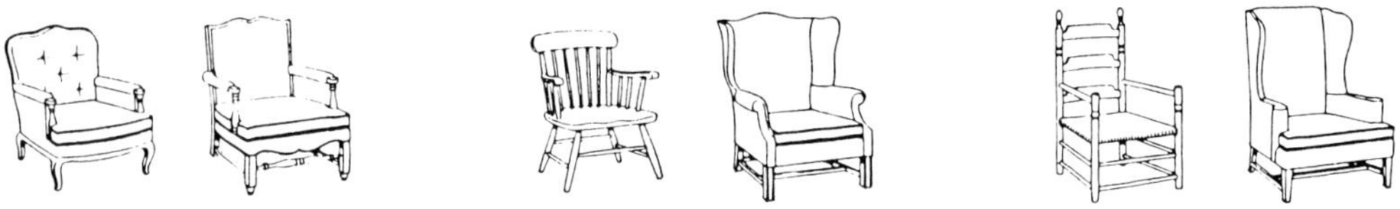

French Provincial | English Provincial | Colonial New England

MODERN FORMAL

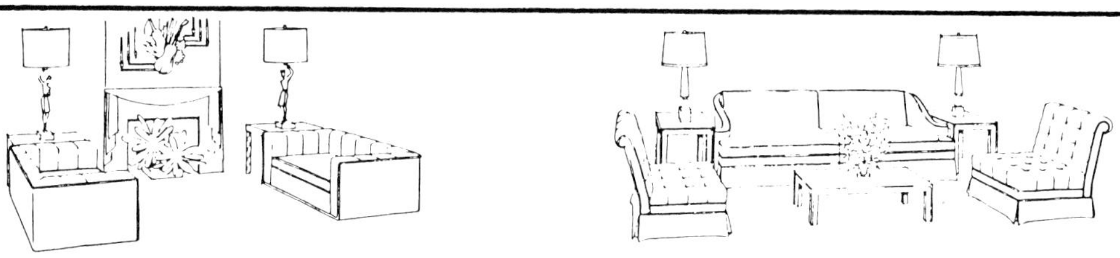

Decorative Modern | Classic Modern

MODERN INFORMAL

Swedish Modern | Mexican Peasant

evolutions in products are alterations ordered by corporations courting the market. Changes in design are what keep people buying a product. Planned obsolescence is built in. Cars, clothes, stoves, and lipsticks go out-of-style, and consumers are prompted to replace them. New designs, then, signal the temporariness of the product as well as advertising the new, up-to-date model. But TV set evolution does not in general follow the rules of planned obsolescence. Instead, the design of the TV set quickly came to focus on unobtrusiveness of the set, flexible viewing arrangements, affordable cost, and good picture quality.

There's real humor in researching the official history of TV design. It reads like one long and often failed attempt to create planned obsolescence so that middle class families would run out and buy a new TV set every year or two the way they did (or wanted to do) with cars. But TV sets were sturdy, and the most significant design innovations occurred in the home around them, not on the set itself. Not for a lack of trying by TV manufacturers. Lying in the gutter of official TV design history are mod and unpopular innovations like zoom TV ('70s) or luxurious wood cabinets ('50s) or giant Advent screens ('70s), TV variations touted in *Newsweek* and analyzed in *Consumer Reports* but rarely seen in anyone's home.

There were only a few key changes in the TV receiver that did make a difference to consumers, and these not at the times when they were invented but at the times when they became accessible and affordable: the portable TV, the color TV, the cable hook-up, and the VCR.

RCA first began to manufacture electronic TV sets in 1939, but domestic production was curtailed during the war, so mass production actually took off only in 1945. The number of U.S. households owning a television rose from about 10,000 in 1946 to 35 million in 1955. Already by 1957 articles were appearing like *Time's* "The Bottom for TV?" which reported on "television set makers plagued by falling sales and big inventories. . . an industry that has long since passed the stage of easy growth." Concentration occurred in the industry, and, by 1957, the number of TV set manufacturers had dropped from over a hundred to thirty-two.

Unlike almost every other product on the market, attempts to establish class differentiation in TV design failed. Early TV advertisements that tried to promote a classist snob appeal seem strangely misguided. One ad, from *Better Homes and Gardens* December 1952, shows a rigidly upper middle class family on Christmas day admiring their new wood-cabinet TV set tied up with a bow. The caption reads, "the best television for the best family in the whole world!" In fact, studies of the time, such as one published by Frank Sweetser in *Public Opinion Quarterly* in Spring 1955, showed that buying patterns for televisions cut across class demarcations.

Nevertheless, in the late '40s and early '50s, the TV set was often designed as a piece of luxury, wood-encased furniture, either as a free-standing console or as part of a cabinet with doors that closed across the screen when not in use. Motorola, for instance, paid elaborate "scientific" attention to how various (almost identical) designs would fit in with different interiors. Every year they published a different chart diagramming which of that year's models would be appropriate in rooms decorated as period formal, modern formal, period informal, or modern informal. (Period televisions were free-standing, perched atop a four-legged wooden base; modern ones were encased in wood cabinets). Elegant or not, these designs were quickly rejected in favor of a table-top model, common from the mid-'50s on, that allowed flexibility in viewing; being able to easily shift the TV's position meant that individual and group viewing could be accommodated. Accordingly, advertisements for TV sets evolved from pitching wood paneled televisions as status symbols to emphasizing picture quality and portability.

In the mid-'50s, Japanese and European manufacturers entered the narrowing U.S. television set market. In response to increased competition and decreased sales, manufacturers produced the "portable" TV set—which did not catch on with consumers until the early '60s. The portable was to be the family's second set, and therefore to double sales. This less bulky television was made possible by the development of the wide-angle picture tube which shortened cabinet depth.

Following (and sometimes competing with) black and white portables, color TV was the next big marketing push. On January 6, 1964, *Newsweek* celebrated: "For 60 minutes on New Year's Day, owners of color TV will have an absolutely unprecedented treat: They will be able to dial any of the three major networks and get color." These were the glory years of the NBC peacock; NBC—whose parent company RCA had invented color TV—had an early almost-monopoly on color broadcasting, a high priority of the network in order to boost RCA color TV sales. RCA claimed half the color TV sales for 1963. In 1964, about 2 million out of the 60 million TV sets owned by U.S. families were color televisions. (Although color TV was invented by RCA in the '50s, it was not commonly affordable until the '60s.) From the mid to late '60s, sales of color TV boomed. GE introduced a portable color TV in 1966. The mid-'60s also saw the introduction of remote control, and additional portable sets often appeared in bedrooms.

The late '60s and early '70s saw a small and ultimately squelched revolution in television content with alternative TV produced on portable equipment brought into homes through cablecasting. Local and national regulations required at least a certain amount of access for local programming. However by the mid-'70s, profit motives had for the most part overridden protections for local programming. Cable had come to the home, but it carried HBO (recycled Hollywood movies) not the voices of community activists. By 1975 VCRs were on the market and alternative TV came to mean a revolution in TV technology, not content.

There have been two major changes in the TV set in the '80s and early '90s. First, as advertising has become more and more ubiquitous, showing up in unexpected, but perhaps not unpredictable, places like children's games, shopping mall walkways, and subways, the TV set has been adapted accordingly to fit a variety of installations—such as banks of televisions in a mall or TVs suspended from subway ceilings. Second, although the home TV has remained surprisingly the same since the mid-'50s, a table top, semi-portable version (with the slight variation that now almost all sets borrow the high tech look of the leading Japanese designs)—what's new in the past decade is that the television has become accessorized. The VCR is now a household fixture. As *Consumer Guide* homily put it in 1988: "Today's televisions are a far cry from the simple sets many of us grew up with. After years of being used solely to receive network television broadcasts, the family TV is being called on to perform tasks undreamed of a few years ago. Television sets can now access cable TV, stereo broadcasts, video cassette recorders (VCRs), video disc players, personal computers, and teletext information systems."

The marketing battles now ongoing are not primarily

in the arena of design innovations. Rather, their outcomes see-saw with the fluctuating tariff relationships between Japan and the U.S. With the upcoming introduction of HDTV (High Definition Television) in the U.S., international trade issues will continue to be paramount.

With the splintering of network program dominance and the individualization of viewing patterns courtesy of accessorized "home entertainment centers," the question remains: have any fundamental and more democratic shifts occurred in either TV programming or viewing patterns? Given my emotional history with television, I hope that the sitcom is never outmoded, and I worry that the MTV format will swallow up TV dramas the way that top ten hits displaced mysteries and plays on radio. So what I want the new technology to deliver is a coexistence of different voices on television including more democratic expression and the ever-satisfying variations on network stand-by formulas. Viewers like me whose television tastes were shaped both by the sitcoms of the fifties and by the community video movement of the late '60s and '70s can take equal amounts of solace and alarm from the top-watched new TV show of Spring 1990: *America's Funniest Home Videos*. This show is determined, in a way, by design changes, the portable cameras, tapes, and VCRs now attached to the sets that allow America's families to record slapstick moments at home. The moments are edited down to micro-moments and broadcast to millions on the network show. This is a highly straight-jacketed (if funny) version of public access.

But has the function of the TV in the home really changed? It is still an emotional free zone, an almost neutral screen (the formulas are so familiar) for group viewing, one of the few activities shared by a family. Viewing has gotten more sophisticated—often, in households that can afford it, TV will be watched while music is playing, a computer game going on, etc. But TV is still the focus of fascination, and it is still an effective lightning rod for family tensions.

Recently, my family had a reunion that lasted a long day and evening. There are more of us now—with the addition of spouses and children—and no charts to predetermine behavior, so it got nerve-wracking at times. Fortunately it was a Sunday and *America's Funniest Home Videos* was on. A couple of my brothers, a sister-in-law, several nephews, a niece, and I retired to the playroom where the TV and VCR were. We switched on the television, sprawled on the floor and the couch, had some laughs, and got out more sinister emotions by describing what we could do to one another (all in the realm of safe violence) to record some slapstick and get ourselves on TV.

This '90s desire for interactive TV—whether it is renting tapes, making your own, plugging in video games, or recording violence to family members and getting it broadcast—seems a fitting step in an intimate history of how we design our spaces, habits, and even our emotions around the television. ■

The Walker Smiths enjoy

Comfortable televiewing in a small room

When the television show is over, you can push the set back inside the cupboard. Louvered doors give the wall interest; a writing desk fits recess below. And the rest of the room is ready for conversation or relaxing as soon as you swing the easy chairs around to face the sofas

Television set pulls out of the cupboards on metal track and faces the sofas. Easy chairs in front of the set swing around for those who like to sit close. The set is placed high, so you can watch a show from every seat in the room. Architect: Arthur Lavagnino

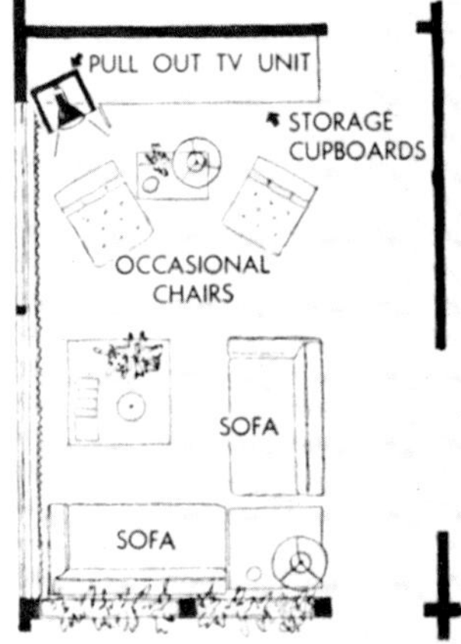

The Smiths have a good arrangement for conversation as well as televiewing

Photographs: de Gennaro

At showtime, you'll want to lounge on one of these cornerwise sofas — across from the television set. Coffee cups and ash trays rest safely on king-sized tables. Casement curtains pull across the window wall when show starts. Helen Logan, decorator

By Barker Childs

You ask: Can a small room offer "ideal seating" for television?

The Walker Smiths say, "Yes." And their room is only 13x18½ feet.

Study these pictures, and you'll see how the Smiths did it. They put the set above eye level so the screen can be seen from across the room. Their two sofas and two upholstered chairs, sensibly mixed with big tables and lamps, fit the space neatly, and are all within easy viewing range of the television set.

Look over carefully the arrangement in the Smiths' study. You may find ideas you can use when planning for televiewing in your home.

BEFORE. This little-used bedroom had long needed redecorating. The center closet was deep enough to hold a television set the family wanted

By Ruth W. Lee

Where do you put your TV set?

In a bedroom, say the Irving Winters of Glencoe, Illinois. You, too, may find it the most practical spot for televiewing

AFTER. Both television and radio fit into the center closet—leaving plenty of hanging space on either side for guests' or out-of-season clothing. In a TV room like this, you can arrange furniture permanently for televiewing. And youngsters can see their favorite shows without disturbing family

Photographs: Nowell Ward & Associates

AFTER. When the closet door is closed, the TV room becomes an upstairs sitting room and guest bedroom. Freshly painted woodwork and new draperies and slipcovers make the room a pleasant place to spend evening. Sofa bed and easy chair are comfortable seating—for reading, talking, televiewing. Agatha Shoenbrun, decorator

Television in the den?

Television doesn't always have to be in the living room. Some families say they definitely like it better in the den. You may, too

Photograph, Yuichi Idaka

Even if your den is small, you can have a television set, radio, and phonograph there—without using floor space. Just put them in the wall. With the set slightly higher than eye level, you can even place low furniture in front of it. Build shelves to fit your record albums, magazines, and best sellers. When you aren't using it to play your favorite records, phonograph slides into shelves and hides behind a door

Suter, Hedrich-Blessing

The Allen Bulleys, of Kenilworth, Illinois, think television in the den is like having a small theater all their own where the show won't bother those with other things to do. At a party, guests who would rather visit or play bridge don't feel hampered by a television program. The Bulleys placed their set in front of a window surrounded by built-in bookshelves. The draperies can be drawn during the show

ANOTHER EXCITING SECURITY SYSTEM FROM GBC!

NOW... YOUR TV CAN SHOW WHO IS AT THE DOOR!

GBC TV SENTRY

MINIATURE TV CAMERA SECURITY/SURVEILLANCE SYSTEM

Connects to any TV set as easily as attaching an indoor antenna.

FOR BUSINESS, THE GBC TV SENTRY...

- Protects personnel and property.
- Keeps you aware of whatever happens in any area of your business that you install a Sentry TV camera...
- All this on your own TV
- Only GBC eliminates the need for monitors

From *Science Illustrated*, July 1949.

Watching TV

Ed Bowes

It's not easy to sit here and write about the TV set. I can't seem to get a large enough or a small enough perspective. I'd like to go and watch my TV. Actually, I'm in the business so I have a bunch of them. I'm not ashamed of all my TVs. Now why would I say that? I'm a little ashamed of the one I watch.

Sometimes I can't look at my hand as I turn my TV on, thinking there's something sharp in my mind that wants taking care of, something unpleasant that I'm not ready to take care of. Something I should do or think about. Then, before I know it, the TV's running. But it doesn't stop me thinking about whatever sharp, unpleasant. . . thing.

I guess I'm not ashamed of my TV. I'm sort of ashamed of my TV watching because it doesn't always work. It doesn't lead directly to something. Of course it doesn't claim to. It just says it will keep giving me more pictures and sounds and that seems like a pretty honorable promise to me. I guess some shows and some networks promise to make me feel better, but I don't believe that and I don't think anyone else does. The stories these shows tell are the same old dumb stories people have always told—there's a problem, it gets defined, then solved. What a vulnerable, old-fashioned concept. Really, that can't be the reason I'm watching all the time.

No, I think the reason's more modern than that. Don't be confused by that idea modern. Don't be too sure you've already dealt with it. And don't just write it off with the Cubists. Modern is one of those big concepts like hunting/gathering, or agriculture, or theism, or the industrial revolution. It's a big gradual adjustment in the way people organize their lives. And one of modernism's particular tenets is that the working descriptions of things or events are not always neat, or well-connected, or linear. Modernism recognizes that some complicated things are indeed complicated, and that that's the best way to understand them—to understand them by their details as part of their whole. Maybe that's what my TV delivers, what I admire and want so much from it—less than fully connected details.

Sure, individual shows try to unify things but they don't do it too well, and I think that TV watchers are getting less and less interested in that unification—just as I think that Einstein, the great scientific modernist, failed at a unifying theory because such unification was antithetical to his basic, modern understanding of things. I doubt that Einstein thought that TV was stupid. I'm sure he had a more useful reaction to it than disdain.

Yeah, I don't watch my TV for the individual shows with their swarmy closures; it's not necessary to make me feel good or fulfilled every few minutes if you want me to watch you.

So what do I have to be ashamed of. I'm OK, I watch TV for its details of picture, word, sound, and event. Because it provides for me the very same type of cognitive stimulus that reading a book does, only in more detail and with more stimulus. Sure, OK now I feel good about watching my TV. I better, I do it for about three hours a day.

Actually, I don't watch TV at all. I read, work, and talk to smart friends; I go to plays. And ever since the invention of TV I feel better about going to the movies. The movies used to be a waste of time, almost an occasion of ignorance, with their second rate little romances, empty successes and noble adventures. But now they're OK, since TV took their place as the mind drug for people who just don't care. Do you think TV has ruined children, made them unable to think right, concentrate long? I really have to get to the TV. I'm not kidding. I have to think a bit. I'm ashamed, but I have to. I'll try to find something out while I do it and report back. . . .

Here's something. I went to turn on my regular TV

in the other room—there's a good show on, but the remote was here on my desk, so I had to come back. Which reminded me (reminding is the way TV works sometimes) that in a recent study, people (that's a cross section of us who use their TVs three hours a day) reported that they enjoyed TV less when they used their remote controls a lot, but that they used their remotes anyway. People work hard. And when they watch TV, pleasure is not their only goal. I'm going to watch now.

I liked it and then I got mad. That happened twice. I fell out of the story a lot. I thought about the actors instead of the characters. I thought about the writers and the people who made the pictures. I thought about things in my life that the sets and the decoration in the sets reminded me of. I had a couple of good laughs. I didn't use my remote because I was resting or thinking so much. Later I saw a story about South Africa that made me cry about a lot of things. But were they real tears, solid laughs, appropriately directed?

Tonight, I'm not ashamed of my TV, or how much I watch it. I think I deserve the time I take, mulling things over while I watch. Ask anybody who's watching TV what they're thinking and they'll be able to tell you. It won't have only to do with the particular picture or story or sound that's on. The experience is not passive at all, unless one defines thinking as passive. I guess one might.

Speaking of thinking, try, for a while, just to think about the TV set, just that encased screen, coffee table book sized. Don't think about what might or might not be on it. Try not to worry about it. It's just a screen, quite unlike anything human beings have ever seen before.

Think about that TV for a week. Think of it sitting there. Keep it turned off and watch it for a few minutes.

Or think of it not being there. I spent my first five years without a TV. I spent years at school without one. Sometimes I get busy and it's almost like not having a TV. I'm not that way now (that busy or removed from TV), and there is a sense of isolation that I miss. Isolation isn't that progressive a feeling. TV replaces that isolation with alienation. In a crowded world, alienation seems more useful than isolation. I think that's progress.

Buy the best TVs you can afford and hook them up to cable until something better comes along. Something better will. TVs will probably get flatter; sometimes bigger and more screen-like, sometimes smaller and more book-like. We may have more control of what we choose to have them display. There will be more detail in their pictures, text, and graphics. We'll get better at using them. We may use them less. Today, *The New York Times* said we already might be. The writer seemed so pleased to say it. If the report was correct, it demonstrates the *Times's* prescience in reporting and analyzing dance more comprehensively than it does TV.

Or, of course, we may use them more. It's hard to predict what the modern world will become. ■

The one and only thing NEW in television!

ZENITH "SPACE-COMMAND" TV TUNER

It answers silent commands from your easy chair...or even from the next room. Turns set on and off, changes stations, mutes sound, shuts off long annoying commercials!

You'll be amazed! There's nothing between you and the television set but space! No wires, no cords, no batteries, no radio control waves. Yet the "SPACE-COMMANDER" control box in your hand carries out your commands from across the room, or even from the next room. Is it magic? How does it work? Well, *see* it yourself...*try* it yourself at your Zenith dealer's. It's like nothing you have ever seen before—anywhere, anytime. And only Zenith has it!

NOTHING BETWEEN YOU AND THE SET BUT SPACE

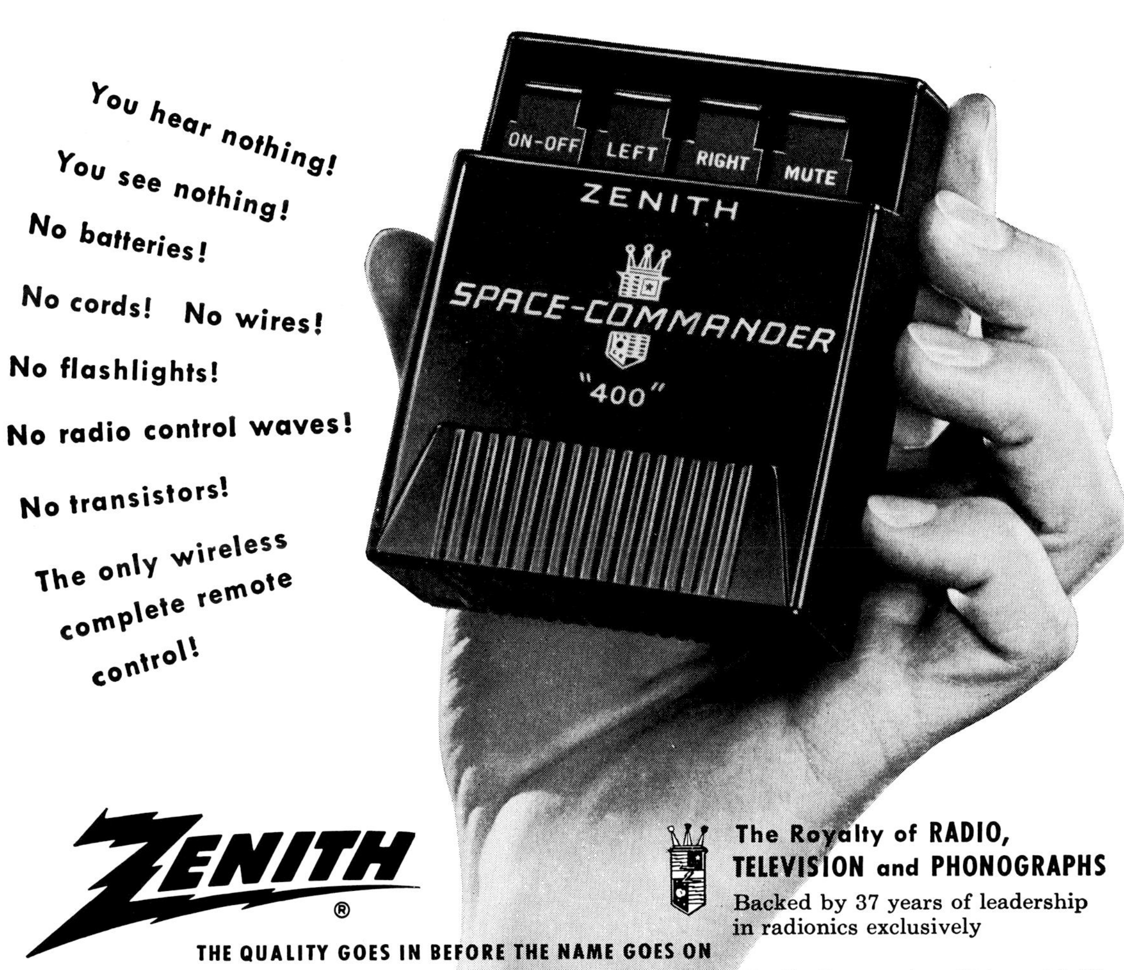

ZENITH®

The Royalty of RADIO, TELEVISION and PHONOGRAPHS

Backed by 37 years of leadership in radionics exclusively

THE QUALITY GOES IN BEFORE THE NAME GOES ON

ALSO MAKERS OF FINE HEARING AIDS • Zenith Radio Corporation, Chicago 39, Ill.

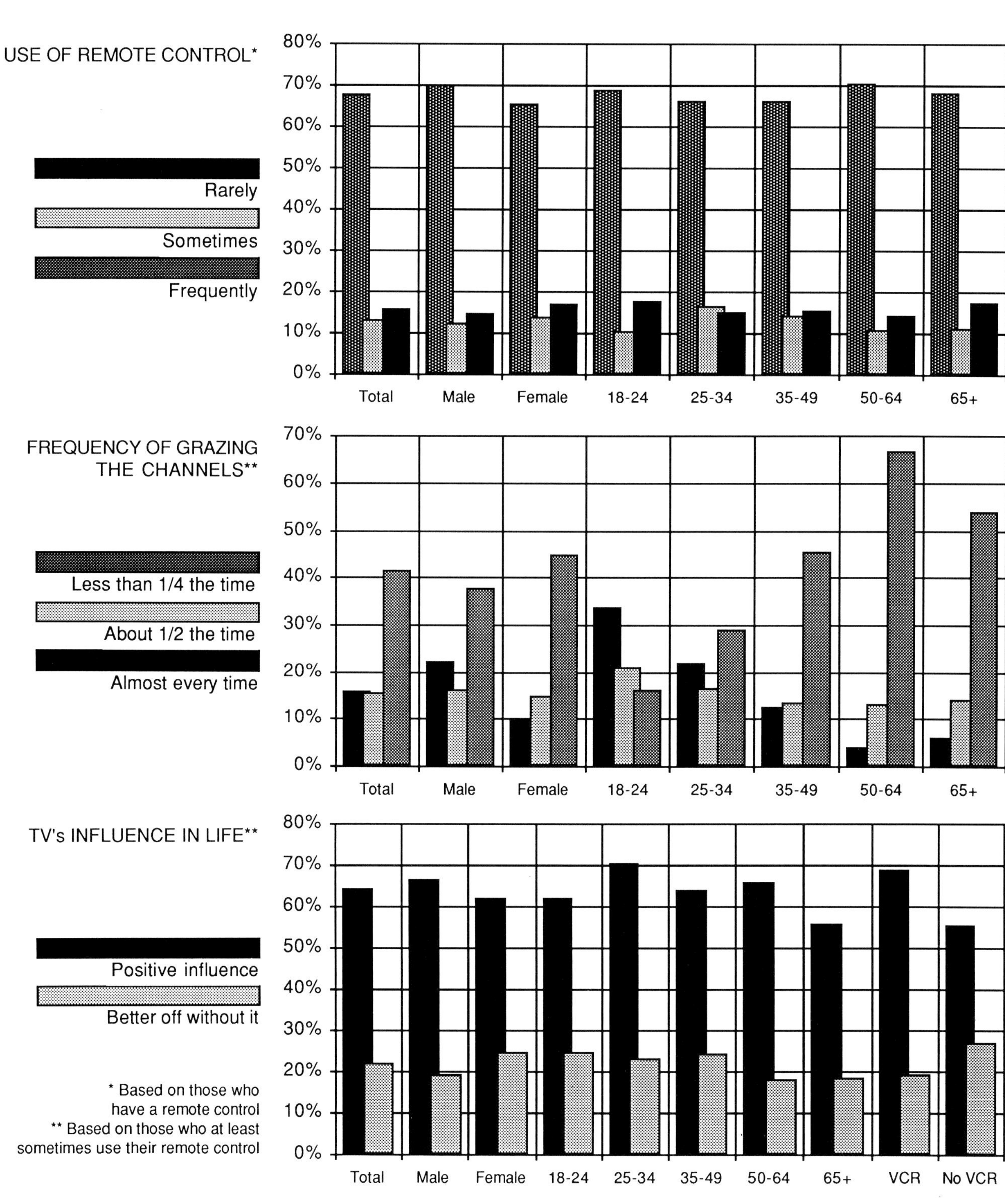

* Based on those who have a remote control
** Based on those who at least sometimes use their remote control

How Americans Watch TV: A Nation of Grazers

Excerpts from a Channels *publication*

In 1989, Channels *magazine conducted a survey of 650 persons nationwide to gain a more thorough understanding of how viewers watch television. They published the results of the survey and commentaries by Merrill Brown, Richard Gilbert, Peter Ainslie, Horst Stipp, Gary Selnow, and David Bollier in an 84 page book. We have selected nine short excerpts from the book.*

The fickle fingers of American television viewers are hovering today over the buttons of more than 70 million remote-control devices, significantly changing the relationship between the audience and the TV set. They flip, zip and zap with frequency, passion and determination, altering viewing patterns, program strategies and even the look, sound and feel of program services. We call the behavior "grazing," and for all the confusion about the subject, there is no getting around the fact that the phenomenon is one of remarkable magnitude. And all indications are, it is here to stay.

*

Two innovations are behind the emergence of this new form of television behavior: the remote control and cable television. Our research shows that 75 percent of TV households have a remote control for either their TV or VCR, and that 54 percent have cable. Some 46 percent have both. It is apparently an irresistible combination—and an adman's and programmer's nightmare: only 16 percent of that group say they rarely use their remote. Of the more than 80 percent who use it sometimes or frequently, almost half of them say they change channels during programs. While viewers 50 years of age or older are less likely to do so, 60 percent of the 18-24 group say they graze. (Even in non-cable homes, we found the remote was used to change channels during programs almost half the time.)

Why do viewers graze? Boredom with what they are watching is the most frequent answer, given by almost 30 percent of those who change channels during a show. But the second most popular response—to ensure that they're not missing a better program elsewhere—amounts to two sides of the same coin. About a quarter of the viewers say they switch to avoid commercials, while 11 percent do so to keep track of more than one program. By combining the first two groups with those who change channels in order to follow more than one program, it is evident that almost two-thirds of the channel-changing during programs is a result of insufficient viewer interest in what they're watching—that is to say, boring programs.

*

Selnow says that observers cannot understand grazing unless TV viewing is considered a merging of two great passions: the nation's love of hardware and the nation's love of television. The American love affair with TV permitted us to sit happily before a small black-and-white set comfortably watching three channels. Now viewers face big-screen color and, with the VCR considered, thousands of choices and the opportunity, through the remote, to survey easily dozens of those options. Hardware and affection have been joined together, he argues.

*

Viewers also exhibited a strong attachment to the medium when we asked about giving up television. More than 40 percent say it would be very difficult for their families to give up TV for a month, and 55 percent say it would be difficult to give it up for six months. Almost two-thirds believe that television has been a positive experience in their lives, while fewer than 25 percent say they'd have been better off without it. And when we asked viewers how much money it would take to get them to stop watching television, more than a quarter said either a million dollars or that they would not stop watching TV for any amount of money.

*

TV viewing has changed greatly since Zenith's Robert Adler invented the remote control and the company introduced it in 1955. Those early devices were far from perfect. The first one was nothing much more that a highly directed flashlight that activated a photo cell in the TV, but buyers found that sunlight and other stray light beams could set off an involuntary round of grazing. Another model, which worked on radio waves, tended to change channels at the neighbor's house as well, and it never really emerged from the laboratory. The next one to go to market worked on ultrasonic tones, but it soon became apparent that it could be triggered by a ringing phone or even a clanking dog chain.

Adler recalls that his boss at Zenith wanted the device in order to eliminate commercials, which he felt were destroying television. The *Channels* survey shows that 34 percent of viewers with remotes do just that: They switch channels at least half the time when a commercial comes on.

Remotes have come a long way since 1955. Ultrasonic tones have been replaced by infrared beams, and there are a variety of such devices on the market, including an amplifier that enables viewers to use the remote without aiming it or even lifting it from their lap. Brand-specific remotes work only for the TV or VCR they come with, but universal remotes can be taught to work with any remote-equipped TV or VCR. Skipping out on commercials is but one of their many skills. Equipped with as many as 50 tiny buttons, they can be used to dim lights, regulate kitchen appliances or thermostats, skip tracks on a CD player or adjust tone controls on an amplifier. The *Channels* survey found that at least 66 million households have remotes, but even that figure doesn't include the simplest of them all: a cable box on a long wire. How many of those are out there, no one knows.

*

Survey results also point up a number of inconsistencies. If Americans are so dissatisfied with television, why do they spend an average of seven hours and five minutes per day watching it as the latest Nielson survey indicates,[1] and why did almost two out of three people tell the *Channels* surveyors that television is a positive influence in their life? One possible explanation is what survey experts call the "halo effect," whereby viewers put on their halos when asked about their viewing habits, but take them off when they actually sit in front of the TV set.

"People are more likely to say they watch and enjoy a PBS documentary than a network sitcom, no matter what their actual habits," explains Frank Walton, president of New York-based Research & Forecasts Inc.

Indeed, a survey conducted last year by the market research firm Edwards Associates of San Diego confirmed that many Americans are embarrassed about their viewing habits. When asked whether they spend "quality time" watching television, 59 percent initially said they did. In the follow-up phone calls, however, 19 percent of them admitted they had been lying.

*

People frequently graze because they are impatient with what they see on their television screen, but others just want the most from the time they spend watching TV. Like more than half her age group, 18-year-old Jeanette Bonilla from Coppers Cove, Texas, likes to watch more than one program at a time. "It's an art to catch just enough of different story lines to follow all of them," she says proudly. "My parents can't take it. I usually end up alone in front of the television."

Sporting events probably offer the most fertile territory for grazing. "Baseball is a half-watching kind of experience," explains Keith Barnes, a project director at a New York research firm. "Remote control is obviously something I've been waiting for all my life. I can watch three games at once and never miss a pitch."

*

However they try to counter grazing, programmers will be hard-pressed to eliminate it, since grazing is not just a vote against commercials or boring shows but a form of entertainment in itself. Along with computers and video games, the remote-control device has helped produce a new, postmodern television viewer, these researchers say, an interactive player who programs his own, constantly changing television menu. Diana Meehan, a communications professor at the University of Southern California, notes that "grazers do not watch television logically the way they read books. Instead, channel switching is a way of making up your own mosaic of images." Like tabloid readers who skim the headlines and may read a few stories, she says, grazers scan channels, stopping here and there when they see a program or commercial that looks interesting. Entertaining though grazing may be, however, Todd Gitlin and other educators worry that it is still a discontinuous, and essentially mindless, intellectual activity. Says he: "Grazing may be better than staring inertly at television. But if watching any one show for an hour is a waste of time, is watching several shows for five minutes apiece better?"

*

The notion of passive viewers was popularized by a book entitled *The Plug In Drug* (Mary Winn, 1977). It professed to expose "television addiction." In a recent forum for scientists and reporters, researcher Prof. Daniel Anderson expressed astonishment over the fact that the "passive viewer" idea was so widespread and popular. He called it a myth that not only lacks scientific research support, but is actually contradicted by research evidence. His conclusions are supported by more general research on the uses of mass media.

*

In sum, the term "grazing" is new, while the behavior is not. It is a kind of selective exposure. There has never been the kind of passive media consumer that is supposedly being replaced by the active grazer. Multiple viewing options and remote controls have not created a new viewer, but they have increased and facilitated the desire to select and create our personal media menu. ■

NOTES

1. Seven hours and five minutes is actually the average number of hours per day a TV set is on in the home, rather than the average number of hours a person watches TV per day.

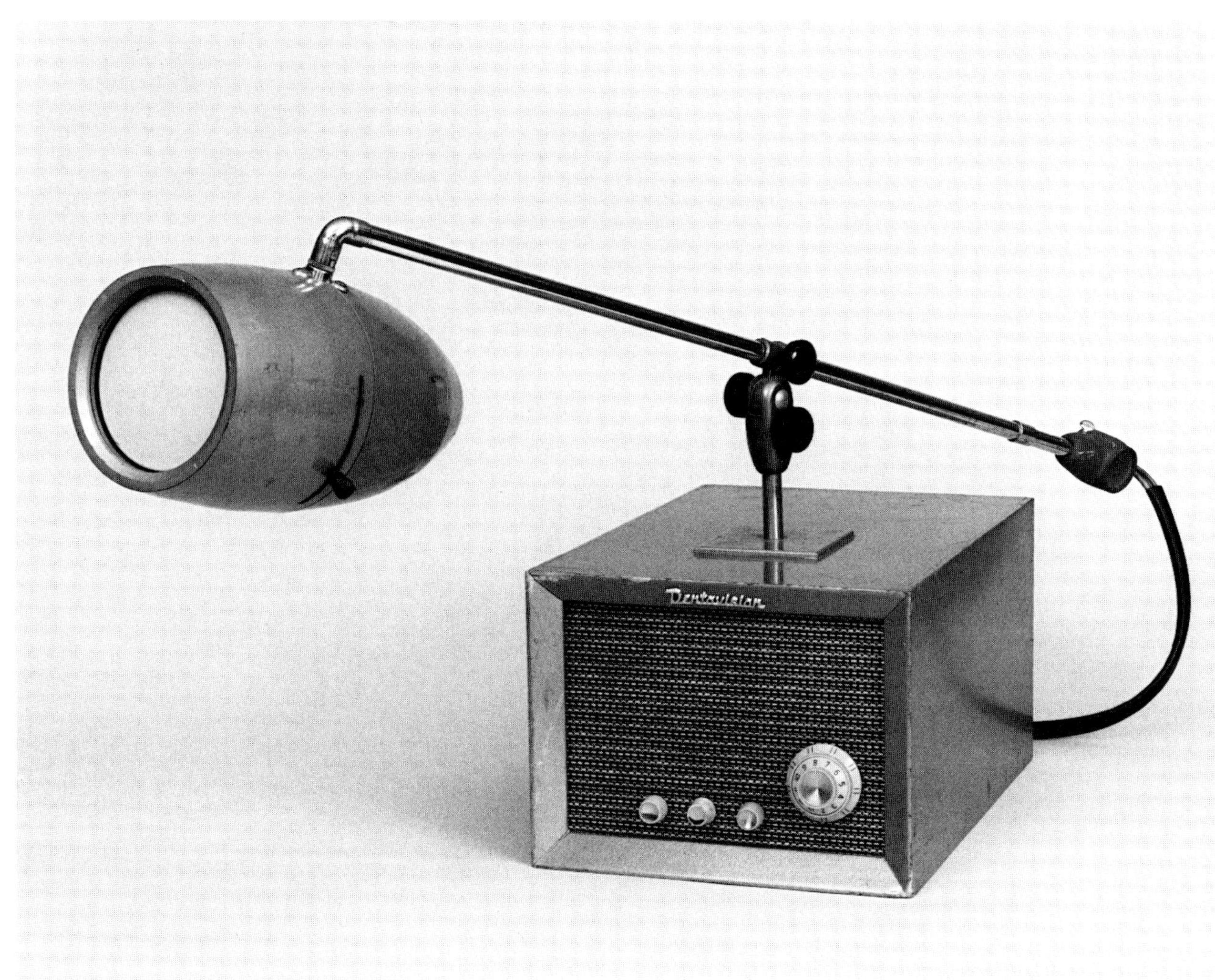

Dentavision, c. 1959. Collection E. Buk, New York, NY.

Publicity photograph for Zenith "Avante" (C4730X), 1972. A 25-inch diagonal console receiver with rosewood color cabinet top, ribbed grill, and frame around the escutcheon provided a sharp contrast to the Bermuda shell white case and broad pedestal base. "Chromacolor" 100 picture tube, new "Titan" 101 chassis and "Super Gold Video Guard" 82-channel tuning were major performance features. (TV reception simulated)

Vol. 3 MARCH 1930 No. 25 25 CENTS MONTHLY

TELEVISION

After Network Television

Les Brown interviewed by Barbara Osborn

Les Brown has reported on the broadcasting business for nearly 40 years as a staff writer for *Variety, Downbeat,* and *The New York Times*. He is author of numerous books on television including *The Business Behind the Box, Keeping Your Eye on Television,* and *Les Brown's Encyclopedia of Television*. From 1980 to 1987, he was Editor-in-Chief of *Channels of Communication,* and is currently Editor-in-Chief of *Television Business International*.

I gather from your column in Channels *over the last year that you are firmly convinced that the days of network dominance in the television business are over. Would you describe what has led to the current situation?*

Technologically speaking, the most significant thing that happened was the satellite. The satellite is a delivery system so much easier and cheaper than the land lines they used to use. With the satellite you can simply bounce the signal in the air and it comes down everywhere. And it's relatively cheap. The networks used to control the industry because they could afford the line charges, which was the only way to get into all the homes in the country.

In 1975 the TV revolution really began. HBO, which was a regional pay TV operation, put their signal on a satellite and had an instant network. Cable systems that had a dish could downlink the signal. As it happened there were only three or four cable systems at the time that had them, but within a year there were hundreds. HBO showed the way to other operators like ESPN, Nickelodeon, MTV, CNN.

Perhaps even more revolutionary than the satellite is the proliferation of remote control tuners. We watch TV differently. We graze over the TV landscape. If I'm looking at something that's not that compelling I wonder what I'm missing someplace else. I can watch four baseball games and two movies at the same time. It becomes à la carte television.

From a business perspective what happened?

Until the early 1980s the networks pretty much had the game to themselves. They were, in effect, monopolies. They had a good thing going because the advertising revenues just grew and grew. In the late 1970s it became a failure-proof business. You could have a flop and still make money, because the demand for airtime for advertising was greater than the supply of space available. But all the networks made a colossal mistake by believing that nothing was going to change. CBS is the prime example, and they nearly destroyed themselves.

So what happened?

Cable diluted network power. Cable expanded rapidly and the networks' share of audience began to erode.

At the same time there was a growth of independent stations. Independent stations, stations without a network affiliation that buy their own programs, were always around. They used network re-runs because they were proven audience getters, and they got high ratings for sports.

But in the early 1980s it began to make more sense to start an independent station because cable would then carry the signal. So independent stations began to burgeon—some 300 new ones in the 1980s. They cut into the network share of audience with programs like *MASH* and *Cosby* which were very successful.

Then an amazing thing happened. Rupert Murdoch bought the Metromedia stations which were a group of independent stations in many of the largest cities in the United States—New York, Chicago, Los Angeles—you couldn't do a network without them. He also bought Twentieth Century Fox Studios and formed a fourth network, the Fox Network. Fox was made possible by the growth of independent stations.

Will the networks die?

I don't know that all of them will. Not in the near future.

The networks were such a dominant force twenty years ago that the Federal Communications Commission restructured the market to let other players in and make for a healthier industry. They adopted the "Financial Interest and Syndication Rule" which prohibits the networks from engaging in domestic syndication (which means they can't sell programs to other domestic stations) and they can't have a financial interest in a show.

But the networks are suffering these days. They're caught in a double spiral: an upward spiral in program costs and a downward spiral in viewing. The networks argue that they are no longer in control of the market. The pressure is on to get rid of the FCC rule. Probably in the next year something will happen—either the FCC will get rid of it or they will amend it. As soon as that happens the networks will move to own some of the programs they are airing. CBS will marry Disney or MCA. NBC will get married to one of the other major studios. Then the networks will control the production, promotion and distribution systems.

What's going on in Europe? How will this affect U.S. TV?

The big market for television programming has always been the United States, but the European television market is expanding. There are two satellite networks over England, and France now has seven channels of television. Of the seven, four are private channels. Spain has just allowed three more along with regional commercial stations. Italy has 500 commercial stations in addition to three commercial networks. So everything is changing. When you add it all up in terms of total homes, Europe is one third larger than the U.S. And now when we include Eastern Europe, we're talking about a market much larger than the U.S. When the networks are allowed to own programs they will want to own programs that they can sell overseas.

What will all this mean in terms of programs?

We'll see more international programs. There are eight countries that matter in business terms: the U.S., Australia, England, Canada, West Germany, France, Japan and Italy. So obviously you play to those countries. Note that four of those countries are English speaking and in Germany, English is a second language.

But if you really want to hit the jackpot, you will still want to make it in the American market. So foreign companies will produce programming with American stars that look American-made. They'll make them in English and then dub them in their own language or shoot them in both languages. It won't end American-style programs. American-style programs set the pace for the world.

What future do you see for direct broadcast satellite?

The American networks are now really looking ahead. ABC is investing in a lot of British plays and musicals. They're targeting the pay-per-view rights. Imagine five years from now, there's a big hit in Britain and everybody wants to see it. The show comes to the U.S. and ABC owns the pay-per-view rights. Opening night is offered to 50 million households for, say, $25. It's a tremendous box office. If a million households buy it, that's $25 million. That's the sort of thing that's going to be done.

In England now, there are two satellite services: Sky

Channel and BSB. BSB is using a high powered satellite. They've used a new dish called the squarial—its a square shaped receiving antenna that you can hang out the window. Because the satellite is high powered you don't need a large dish and it's not as ugly as the parabolic ones. That could be revolutionary. Certainly when people have those kinds of dishes, then companies will broadcast their own programs rather than selling them to a network.

Is free TV in jeopardy? Will we see the end of advertising supported television?

I don't think so. The real danger to advertising supported TV has been the poverty of the programs. Pay television is a factor up to a point. But unless it's something I really want, why would I pay for it? Advertising supported TV is here to stay.

But program forms have to change radically. Everybody is concerned about the remote control because so few programs are really zap proof. You can watch five minutes of a sit com and you know exactly what's going to happen. A movie that's only moderately interesting gets zapped.

What doesn't get zapped are serials, because you get caught up in the story. You want to see what happens. Americans do soap operas which never end, but the Latin Americans are the real masters of the serial. They make novellas—novels for television. These shows have an ending—it may not come for two years—but they have an ending.

Artistic programs are also relatively zap proof. They don't get big audiences but chances are if you tune in the ballet you'll watch the whole thing. For the same reason documentaries are relatively zap proof. But a lot of these ditzy television programs are not going to fly; *Gilligan's Island*, for instance, is just not compelling enough. I really believe that television programs will probably get better.

Do you have a sense of how many channels can survive in a diversified marketplace?

How many magazines are there? There are magazines like *TV Guide* with a circulation of 30 million readers and there are magazines like *Channels* with 27,000. They both survive. It depends on what you're doing. I can imagine a channel that plays a double feature movie every day, and they won't expect you to watch every day.

How large a role will cable and the multi-service operators (MSOs) place in the future?

The MSOs, the large cable operators, will continue to consolidate. We started out with a huge number of locally owned cable companies, but then they all started cashing out and selling to larger companies. I guess there will be ten maybe five MSOs in the next decade.

Whether cable will be as good a business depends on a number of things. One of the things is re-regulation. Every time a new program service goes on the cable system they charge the cable operator for it. The cable operator then passes the cost on to the consumer. With cable deregulation the rates rose continuously and people have complained. The rationale for deregulation was that the public would pay less but in this deregulation the public paid more.

There's hell to pay right now. There's a lot of sentiment to re-regulate cable. Cable became attractive to buy when you could raise the rates as you wished. It may not be as attractive otherwise.

I don't think cable wants to expand to more and more channels anyway. They're going to start making money selling advertising. Right now the total advertising budget for cable is about 1 billion a year compared to 27 billion on network TV. The expansion possibilities are tremendous. But if you keep adding channels you dilute the audience. What you want to do is capture the audience with the fewest number of channels and then lay on the advertising—the fewer channels, the more advertising at higher prices.

Is the establishment of enormous media companies something we're liable to see across the industry? Will this ultimately limit the diversity of programming?

If you compare what we see now to what we used to see from three networks, TV is more diverse. It could be argued that there are more companies now than there were ten years ago. They're going to get bigger and bigger and more and more international. The U.S. will not be as insular a country as we've been.

On the other hand, look what happened to department stores. These great institutions are going under. In their place are these huge shopping mall complexes which are, in effect, department stores. Maybe that's the future of cable.

What will happen with movies?

In the year 2005, theaters will have satellite dishes, and will take down high definition signals from satellite. You will see movies in high definition television. The technology of high definition lends itself better to distribution of movies than distribution of television right now because you have to re-equip all the TV studios and all the homes. High definition also looks better on the big screen than on the small TV screen.

Will computers become televisions?

Maybe. The computer screen is similar to the TV screen. Videotext is starting to be used for information retrieval. It's also entirely possible that you could have a movie on a disc and pop it into your computer.

But even if we watch a film on our computer, television and video are two distinct experiences. Television programs are seen in a flow. When you pop in a video it's not a flow. There are no other channels. It's a different experience.

So it's possible that in the relatively near future we'll be using our computer as a TV. We'll pop in a movie, have a squarial outside the window and receive 50-100 channels via fiber optic cable.

Technologically, it's certainly possible. The only safe prediction to make is that change will continue well into the next century.

Do you have any last thoughts about this particular period in television history?

This is an interesting period because there are so many new things. More TV is not necessarily better, but it's certainly more democratic. Ten years ago, you couldn't make television unless CBS said you could make television. All over Europe now you see the growth of independent production companies. It has taken the power away from the small clutch of people who decided what would go on the air. A lot of shows are being created that wouldn't have been created for the networks. I like the idea that in a highly fragmented market every kind of thing can find an audience.

But I will always remember fondly the three big networks and how we were in their thrall, how there was one set in the household and everybody watched it. The three networks we loved to hate. We'll never see them again. ■

Mitsubishi VS-521UD Projection Television, 1988.

Wolf Vostell. Scene from the performance, *TV-Décollage* at the YAM festival (HAPPENINGS), organized by the Smolin Gallery in New Brunswick, New Jersey, May 19, 1963. Photograph © 1963 by Peter Moore.

The Anti-TV Set

John G. Hanhardt

This essay is a reflection on how artists first appropriated the television set into their artmaking. Two artists, Nam June Paik and Wolf Vostell, are among the very first, and certainly the best known, to incorporate the TV set as an icon and object into their work. The projects examined here are Nam June Paik's one-artist exhibition, "Exposition of Experimental Television," held in March 1963 at Galerie Parnass in Wuppertal, West Germany, and two exhibitions by Wolf Vostell—a performance/happening at the Yam Festival in New Brunswick, New Jersey and his concurrent one-artist show at the Smolin Gallery in New York in May 1963. Both of these exhibition/events are closely related to these artists' earlier work in performance art and to the Fluxus movement. They were deliberate efforts on the part of Paik and Vostell to transform the institution of television through the destruction of the TV set.

Throughout the history of artists' video there has been a desire, expressed both implicitly and explicitly in writings and artwork, to rethink the institution of television by re-examining our assumptions about programming and the use of the TV set, and transform and appropriate the medium by fashioning alternative programs, circuits of distribution, and exhibitions of the apparatus itself. The insight expressed in these early artists' projects was to rethink, refashion, our perception of the medium's possibilities by confronting both its identity as a pervasive piece of household furniture and its ideology as a distribution system of words and images. These first efforts began where television begins for most people, namely with that technology of everyday life, the TV set in the home.

The art world in the late 1950s and early 1960s was witnessing a period of dramatic change. It was a time when artists rejected the idea of art as both exclusively defined by painting and determined by the lone poetic genius of the artist seeking existential insight through painting abstract expressionist canvases. Certainly this was not the first time in twentieth century modernism that movements coalesced around the idea of rejecting traditional high art definitions and categories of what constitutes artmaking; nor was it the first movement to look beyond the figure of the artist as single creative genius. This was a multiplicity of movements—happenings, Fluxus, new dance, minimalism, performance art, music, avant-garde film—that were each, through a variety of media and materials, returning to the ordinary details of everyday life and retrieving as the material for creative expression the very stuff and experiences of quotidian existence.

At the same time these artists were turning to the materials of popular and everyday culture, American and European society was being rapidly altered by television as a form of popular entertainment and information delivery consolidated by monopolistic and state-run systems. The impact of television was similar to the way in which the cinema and the movie-going experience transformed the public's view of the world in the early part of the century. With television, people began to turn to an audio-visual device that was to construct a new public mythology alongside the movies. Television as a new medium delivering new images for daily consumption into the home was perceived as a threat to the movies because of its convenience, pervasiveness and low cost. The impact of television on popular and consumer culture was and remains enormous. This period, which saw the final victory of television as an industry and as a permanent fixture in both the private space of the home and the consciousness of the individual, was the same period in which artists turned their efforts toward reconnecting artmaking to everyday existence as a non-elitist activity and retrieving the details of daily life as the

Nam June Paik. Partial installation view of the exhibition, "Exposition of Music—Electronic Television" at the Galerie Parnass in Wuppertal, West Germany, March 1963.

Nam June Paik. *Distorted TV*, 1963. Collection Dieter Rosenkranz, Wuppertal, West Germany.

raw material for art. The effort to remake the consumer image in Pop Art, to intervene in the processes and institutions of life in happenings, and Fluxus' creation of an alternative culture all occurred as television further colonized the public consciousness.

For artists working in the early 1960s, before the portable video camera and player were introduced by Sony into the U.S. market in 1965, the medium of television was completely mediated by the TV set. The iconicity of its programs and their perceptions were reinforced by the position of the television set in the home. The first artists to think about the medium recognized the TV set as the emblem or container of the medium, and thus saw a need to first deconstruct its authority as an object in order to ultimately create and work within its system.

It is this profound insight into television, not as a found object to be recontextualized as art, but as an icon to be broken of its authority and rebuilt out of its own parts, that distinguishes the work of Nam June Paik and Wolf Vostell, two artists in the Fluxus movement, a neo-Dada alliance of artists in New York and Europe which followed the call for a new art practice issued by its founder George Maciunas. Fluxus was an anarchic band of iconoclastic individuals who avoided the trap of mainstream culture and society as free-thinking individuals producing manifestos, performances, objects, texts, imaginary institutions, and projects all of which denied the status and stasis of culture.

In 1963 Vostell and Paik each had exhibitions in which the television set played a central role. Paik was a performance artist whose earlier projects involved the destruction of traditional musical instruments, in particular, pianos and violins. The centerpiece of Paik's exhibition, "Exposition of Experimental Television" in March at the Galerie Parnass in Wuppertal, West Germany, was a room filled with thirteen "prepared" televisions. Television sets were scattered about the room with their exteriors scratched and marked and the screens showing either distorted television programs or abstract line and wave patterns created by the direct manipulation of the mechanism of the set itself. Paik's show, with a bloody bull's head suspended above the entry way and mannequins, experimental musicmakers, and prepared pianos scattered about the space, was a shock and assault on the sensibilities of its German middle-class viewers. The link between the prepared piano and the television in Paik's work lies in television's replacement of the piano as the most popular medium in the home for entertainment. Paik's appropriation of the television set is also an extension of his interest in altering our perception of traditional, standardized cultural forms.

Paik's defacement of the piano changed the look and sound of the instrument: its altered keyboard and strings, covered with doll's heads, barbed wire, and other debris of the culture, created unconventional, anti-classical sounds. Paik desanctified the piano and removed it from the pedestal of high culture. As in "One for Violin Solo," in which he destroyed a violin in a single gesture, thus producing its final sound as his concert, Paik transformed the piano as instrument and cultural artifact. In an extension of this strategy, Paik removed the television set from its position within the home and stripped it of its signifiers and traditional meanings as an object.

"I utilized intensely the live transmission of normal program, which is the most variable optical and semantical event, in the 1960s. . . 13 sets suffered 13 sorts of variation in their VIDEO - HORIZONTAL - VERTICAL units. I am proud to say that all 13 sets actually changed their inner circuits. No two sets had the same kind of technical operation. Not one is the simple blur, which occurs when you turn the vertical and horizontal control-button at home."

As with his prepared pianos, Paik denied television its standardized content and appearance by renovating the set's interior and exterior, thus transforming each standardized instrument into his own individual instrument. When Paik began in the late 1960s to produce videotapes he still adhered to his interest in electronic images as an extension of his first involvement in electronic music and performance instruments. Paik's long career and seminal position in the history of artist's television/video sprang from his first prepared televisions at the Gallerie Parnass.

Wolf Vostell's early performance projects were actions designed for public spaces; he was also the creator of the publication "décollage," which used the mass media format of the newspaper to create a new form of anti-newspaper. Vostell's installation at the Smolin Gallery in New York, May 1963, is directly related to Paik's prepared televisions. Vostell employed magnets and manipulated the television so as to transform the received signal, and fashion his own electronic images. Vostell focused on the television set as an object by grouping televisions together with office furniture in the gallery, thus reflecting on the television as the conduit of information and information processing. Like Paik, Vostell manipulated the image to offer the concept of the individual creating his own programming and fashioning a commentary on the TV set and its content. As in his "décollage" newspaper made up of pieces of paper, magazine clippings, and artists' statements, Vostell employed a boldly improvisational layout and design in his prepared television sets.

The Yam Festival event organized by Alan Kaprow, George Brecht, and Robert Watts took place on George Segal's farm in New Brunswick, New Jersey, in May 1963. It featured a "TV Dé-collage" event by Wolf Vostell. Vostell's performance began with a television set covered with barbed wire, its screen reframed by a picture frame, and with a music stand placed in front of it. In the performance, the TV set was first carried outside; Vostell then used a jackhammer and a shovel to dig a hole in which the body of the TV was buried. Vostell's action was a symbolic effort to eliminate television and bury its physical body. This anti-ceremony both acknowledged and denied television through its primary and most recognizable form, the TV set. Vostell's text for the event is a graphic set of instructions to engage and transform the TV set.

Nam June Paik and Wolf Vostell influenced each other as co-participants in a large international community of artists. Paik went on to shape video's history as its most important and influential artist. Vostell, although he did not play a continuing role in video art, demonstrated in this and related projects, especially his performances and Vietnam pieces from the 1970s, a social and political desire to understand the authority and ideological controls apparent in contemporary history. Both artists understood that television begins in the public's mind with the authority of the TV set itself as a conduit for the programs unfolding daily on the screen. Paik and Vostell's efforts to demystify the public discourse of television as well as the practice of artmaking constituted a radical step in the history of video and art. By confronting the instrumentality of television, their work constituted both a materialist critique and an anti-high art activity. They confronted the individual with the detritus of consumer culture in an effort to understand art as a living presence, not as a dead tradition or unquestioned commodity of the marketplace. ■

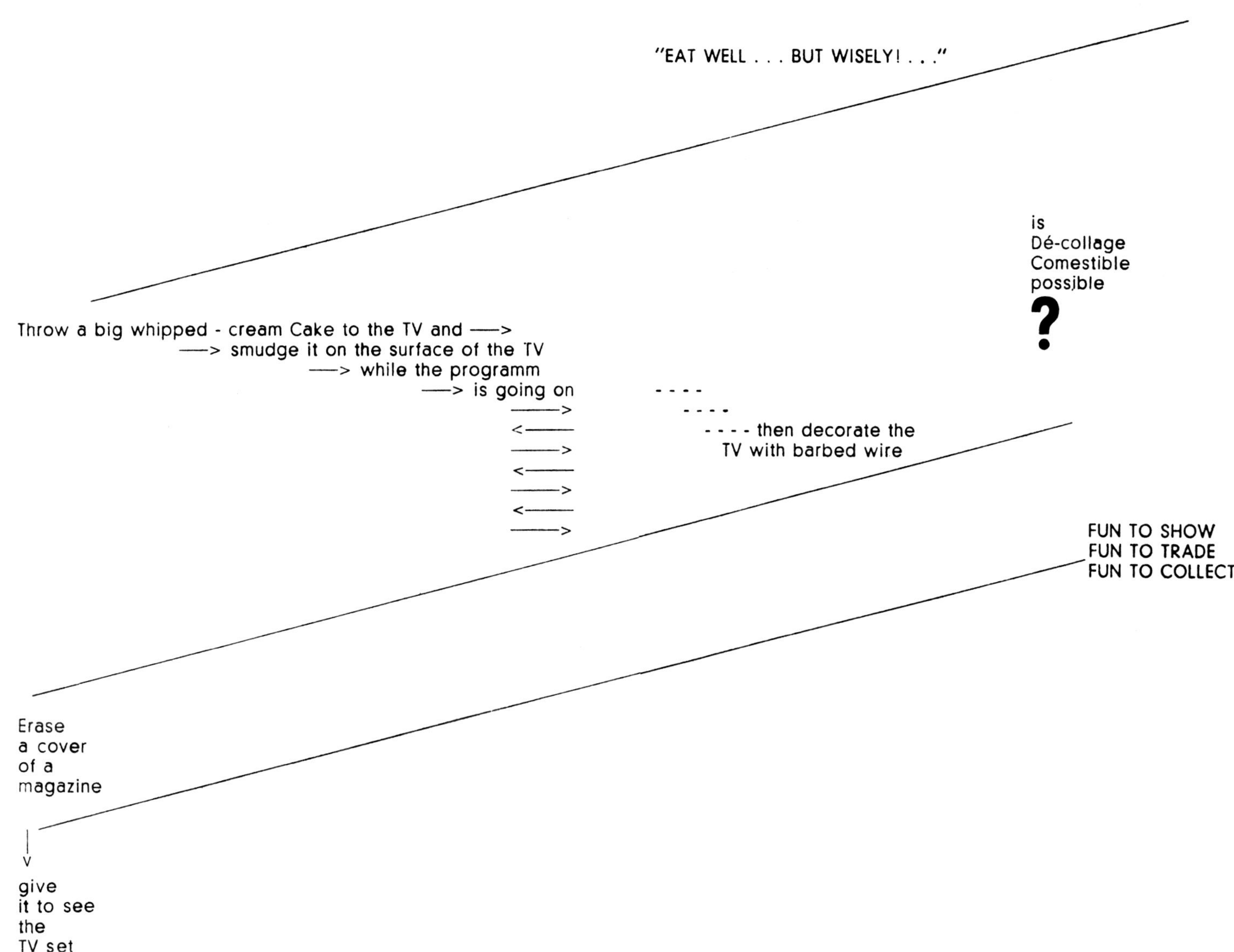

Wolf Vostell. Text for the performance, *TV-Décollage,* 1963.

THE HOUSE THAT WATCHES BACK

ARCHITECT: BROOKS KENDALL SLOCUM
GRAPHIC DESIGN: CHITA CONTRERAS
SURVEILLANCE: JULIA SCHER, (718) 855-7089

GLOBAL OPERATIONS: RESTRAINT SYSTEMS/QUARANTINE BLOCKS/MULTIPLE I.D. INDEXING/DRIVER ESCORTS/COMPUTERIZED MOBILE SURVEILLANCE.

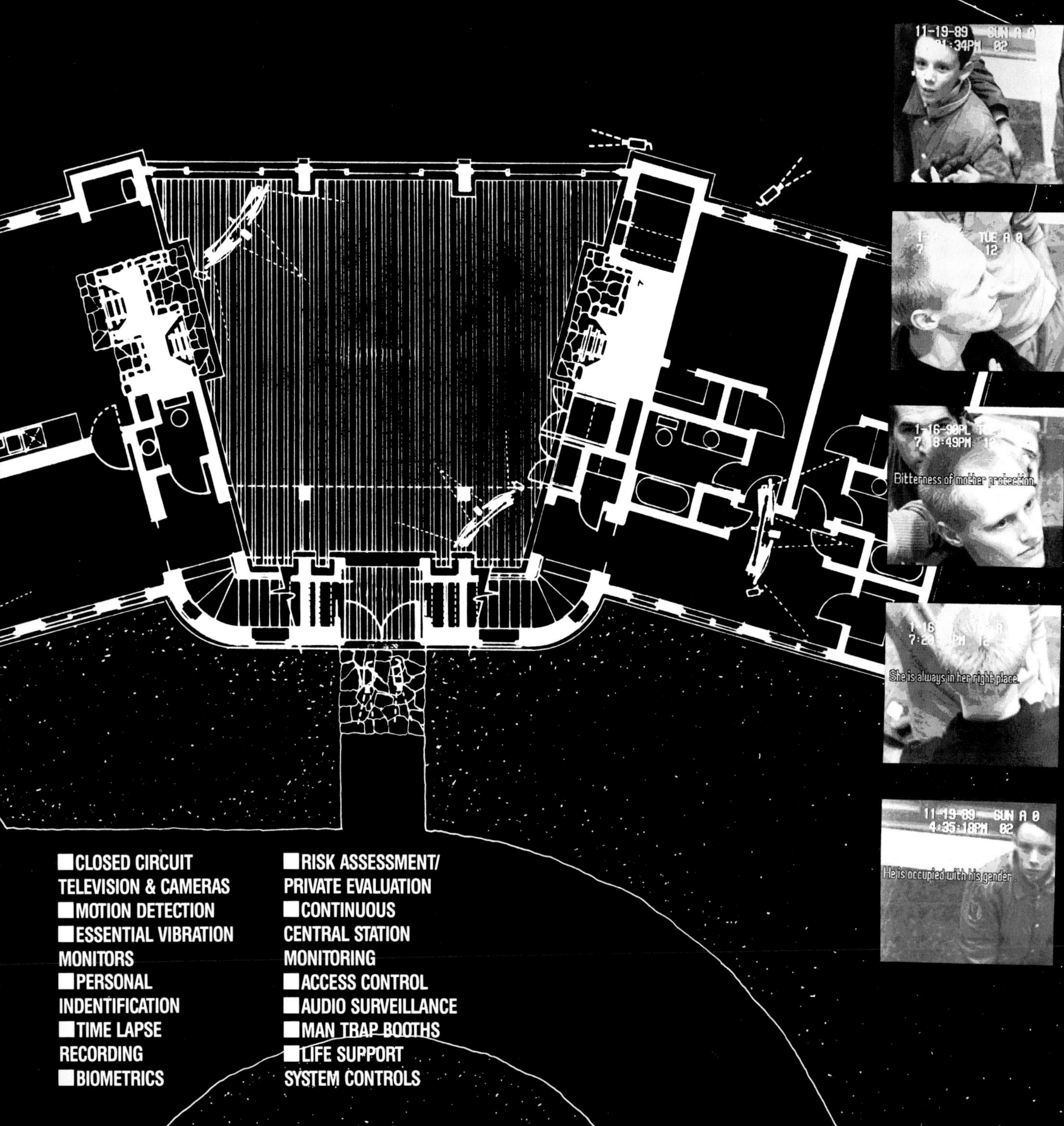

Security By Julia

Member National Burglar and Fire Alarm Association, Inc.

Member Metropolitan Burglar & Fire Alarm Association

OUR
VACATION
RCA
SPACE
T-120 SHG

America, America, This is You: The Camcorder and the Red, White and Blue

Alice H. Yang

You're the red, white and blue, you're the funny things you do. America, America, this is you.
—Theme song of *America's Funniest Home Videos,* 1990

In recent years the camcorder has surfaced as a consumer gadget of surprising popularity. By some accounts, it has already reached the hands of nearly eight million Americans, and estimates are that this number will quintuple by the mid-'90s.[1] While market statistics are a clear measure of the camcorder's ubiquity, other indicators suggest the extent of its cultural impact.

Among these is commercial broadcast television itself, which features camcorder technology as the central device of what has become a hit television show. *America's Funniest Home Videos,* as the program is called, uses footage solicited directly from the home camcorder buff. Screened by the producers and organized loosely around such themes as babies, pets and basketball, the tapes are broadcast with a running commentary by a wise-cracking host. With its weekly fare of antics and pratfalls, the show is a remarkable phenomenon inasmuch as it identifies a brand of national humor at its lowest common denominator.

As *America's Funniest Home Videos* reached the top of the ratings, a number of articles appeared in the press which sought to explain the program's appeal. Reviewing it in *Newsweek,* one journalist wondered, "And why not? A concept so elemental, so economical, so *democratic*, seems ideally in sync with the populist, no-frills '90s. Nor [sic] has any piece of entertainment so successfully exploited our most basic biological urge, which is, of course, to get on TV."[2] It would seem from this one commentary that *America's Funniest Home Videos* is emblematic of the political (democratic/populist), economic (no-frills/economical), and even somatic temper of the moment. Even more telling, it would appear from the *Newsweek* reporter's comments that the camcorder is helping to reassert certain democratic truths.

This theme of democracy has a strangely familiar ring, if only because it has been raised before throughout television's history. In the euphoric tone characteristic of early commentaries on television, one writer in 1946 noted: "Today we stand poised at the threshold of a future for television that no one can begin to comprehend fully. . . We do know, however, that the outside world can be brought into the home and thus one of mankind's long-standing ambitions has been achieved."[3] By bringing a range of entertainment, information, and education to homes across the country, television was regarded as a source of affordable knowledge and a conduit to new horizons which was available to all. "Just as the printing press democratized learning," critic Daniel Boorstin later wrote, "so the television set has democratized experience."[4]

Although this view has had its champions, television has also and repeatedly been criticized for failing to fulfill its democratic promise. As many have noted, market forces have maintained such a stranglehold over the structure and content of television that the much-lauded opportunity it was said to represent for equal access has been undermined by its failure to allow for equal participation in the production and distribution of televisual information. As a result, a handful of national networks have monopolized the airwaves and produced a bland diet for mass consumption. Skepticism about the neutrality of a medium, which has allied itself with advertisers, has fed the anxiety that television is a one-way "window on the world," a medium not of information but of its control.

This critique of television, among many others, has provoked attempts to institute alternative channels and to engender in the viewer a more discriminating relationship to television. The late '60s and '70s, in particular, witnessed a flurry of interventionist activities which were propelled in part by the introduction of affordable video technology. Most notably, Sony's half-inch video *Portapak,* a precursor of the contemporary camcorder first marketed in 1965, provided media activists with an accessible means to produce their own programs. These efforts were further encouraged by the establishment in the late '60s of both public-sponsored and cable television, which became the forum for alternative programming.

Michael Shamberg's book *Guerilla Television,* published in 1971, gave voice to the dramatic impulse driving these activities. Video was viewed as "the means to 'decentralize' television so that a Whitmanesque democracy of ideas, opinions, and cultural expressions—made both by and for the people—could then be 'narrowcast' on cable television."[5] Thus the rights of equal access, participation, as well as self-expression—key ingredients of democracy—ironically resurfaced to challenge television. The debate found its language in a particular American ethos, as attempts were made to negotiate the slippery terrain between private and public forms which was being reshaped by the telecommunications explosion.

Despite the enthusiasm and achievement of its proponents, "guerilla television" largely failed to infiltrate the power structure of American commercial television. Shown infrequently on public stations and a handful of cable channels, programs made by media activists languished in the margins of mainstream culture. Television has proved to be resistant to a true "Whitmanesque democracy," as major networks continue to dominate the airwaves and serve as the model for most cable channels.

The arrival of the camcorder, an apparatus even more mobile and affordable than the *Portapak,* would seem to revive hopes for television's democratic potential. Compact and simple to operate, the camcorder is an accessible device for moving image-making, whose footage has the added potential of feeding directly into TV (both on a literal level, through the VCR and, on a metaphorical level, through the channels of broadcast television.) It would seem, judging from *America's Funniest Home Videos,* that the tools of television, previously available only to the industry and a select number of activists and artists, have been handed to the people. Does this signal, therefore, the coming of television made both "by and for the people"?

The structure and content of *America's Funniest Home Videos* would almost lead us to believe so. For thirty minutes a week on national television, the product of camcorders—home videos—occupy center stage. "The whole idea of the show," in the words of the producer, "is to have America produce it for us."[6] The formula is simple enough: footage is solicited from the viewers and transmitted back to them reflexively, as a kind of mirror image. The democratic motif is further enhanced by the presence of a live audience which is invited to exercise its right of choice by selecting the winner of a weekly $10,000 prize. Remember, democracy is either about the exercise of real choice by real people, or it is nothing at all.

As much as one might believe in this participatory structure, it is finally exposed as one of the program's conceits, which undercuts the very principle of democracy that *America's Funniest Home Videos* seemingly promotes. Home footage is, after all, only the show's raw material. Utilizing voice-overs and other editorial techniques, the producers carefully orchestrate the tapes to elicit the viewer's reactions and guffaws. As a result, *America's Funniest Home Videos* offers a narrow portrait of the American people, one which reinforces domesticity and the idea of the nuclear family as the All-American ideal. Not surprisingly, the paradigm of the white, middle class family is key. And, to quote *Newsweek,* why not? With a starting price of $600, the current costs of camcorder technology necessarily limits its "democratic" reach.

To describe *America's Funniest Home Videos* as democratic, as the press has done, is to mistake for real the shill game of power which television is so adept at. Camcorder videos, presented on the program as a mechanism for viewer participation, are little more than devices that mask the control that belongs to the show's producers. If there has been a recapitulation of the theme of democracy in television, perhaps it is because camcorder technology has lent itself to the democratic rhetoric at stake.

Of its many features, spontaneity and immediacy are the trump cards of camcorder technology. So portable and easy to operate that it has become almost foolproof, the camcorder is the perfect instrument for capturing life's transitory moments. With a flick of a switch, you can record just about anything, anywhere, and at anytime. It is, in fact, these very qualities—at the heart of *America's Funniest Home Video's* "caught-in-the-act charm"[7]—that has led to much of the confusion between democracy and its illusion. Since the footage shown on the program is supposed to document the unforeseen and unplanned, is it not therefore the direct and unmediated expression of the people? Is it not responsive to the call for democracy?

The truth value of home videos, however, like photography, is clearly contingent on how and why these images are made and transmitted. It is not so much a question of the home video's inauthenticity, then, as it is a question of selectivity. Mediating conditions continue to determine what we see and the meanings we attach to them. Conveniently, the clips on *America's Funniest Home Videos* contribute to a complacent, lily-white vision of democratic American society.

In other, more subtle ways, broadcast television has taken advantage of home videos as a proof of spontaneity. Many news programs, for example, incorporate amateur, on-the-scene clips of cyclones, riots and airplane crashes to report on breaking stories. One such clip, taken by a tourist, provided a devastating first glimpse on TV news of the earthquake which hit California in 1989. This journalistic device has become so pervasive that the Cable News Network (CNN), as well as some local stations, now flashes a "video hot line" number at the close of their newscasts. CNN has, in addition, developed a program, called *Newshound,* devoted entirely to the broadcast of newsworthy amateur clips.

The insertion of home videos into the news format indicates that camcorder technology, more and more, is superseding the documentary function once reserved for photography. Amateur clips are a persuasive journalistic tool insofar as they emphasize the factual nature, and objectivity, of the news. What we see with our own eyes is equated with the all-encompassing "truth". With its moving and speaking subjects, home videos seem so sensuously "live" that they have, in fact, come to underscore the "real" more convincingly than photography ever could.

Camcorder technology thus heightens the illusory appeal of television, by blurring our already shaky perception of the "real". The ironies are not lost on the producers of *America's Funniest Home Videos,* who go to great lengths, so they claim, to screen the tapes for signs of fakery—"a definite no-no for a show so dependent on an aura of spontaneity." As one staff member of the program explained, "most of these setups are direct copies of things we've already aired.

People [will] say, OK that's what it takes to get on the show."[8] What is lost in this circulation of images made "by and for the people" (or is it for TV?) is the critical distinction between democracy and its semblance, a distinction which dissolves in an ambiguous, nether region where the patently false is indistinguishable from the evidently real.

And so, all too fluidly, broadcast television has coopted camcorder technology to reinforce its representation of authenticity and at once implicated us in its perpetuation. We are drawn back to the authority of television's images with an ease that belies the slant of its rhetorical framework. But if the apparent "authenticity" of home videos, so much more direct and palpable than photography, can help to legitimize the values endorsed by broadcast television, so perhaps it can also service other arguments, other ideologies.

With its attribute of spontaneity, camcorders have often been effectively deployed, for example, to capture volatile political and social events. During a 1988 demonstration at Tompkins Square Park in New York City, a participant recorded a shocking instance of police brutality. The tape, shown on TV news, was instrumental in garnering support for the protesters. Even more dramatically, when protests brewed in Czechoslovakia recently, students were able to use camcorders to secretly document oppositional activities which the state-run TV station had refused to cover. The distribution of these tapes, played on VCRs throughout the country, stimulated widespread participation in the protest movement.

The case of Czechoslovakia points to an alternative use of camcorder technology that is in conscious opposition to broadcast TV. In that political context, home videos helped to rally support for a brand of democracy resistant to the status quo, a democracy that vividly contrasts with the version promoted by *America's Funniest Home Videos*. As use of camcorder technology spreads and our TV sets become, in turn, ever more "user-friendly," the question is not so much how we can achieve an elusive democracy through television, but how we are to contest its strategies of seduction. ■

NOTES

1. Cited in Harry F. Waters and Lynda Wright, "Revenge of the Couch Potatoes," *Newsweek*, March 5, 1990, p. 54.
2. Waters and Wright, Op. Cit., p. 54.
3. Thomas H. Hutchinson, *Here is Television, Your Window on the World*, New York: Hastings House, 1946, p. ix. Quoted in Lynn Spigel, "Installing the Television Set: Popular Discourses on Television and Domestic Space, 1948-1955," *camera obscura*, no. 16, January 1988, p. 15.
4. Daniel Boorstin, "TV's Impact on Society," *Life*, vol. 71, no. 11, September 10, 1971, p. 36.
5. This quote is a critic's paraphrase of Shamberg's ideas. Deirdre Boyle, "Guerilla Television," in *Transmission*, edited by Peter D'Agostino, Tanam Press, 1985, p. 204.
6. Karen Grigsby Bates, "The Bride is, Er, Excused," *Time Magazine*, March 5, 1990, p. 53.
7. Ibid., p. 53.
8. Waters and Wright, Op. Cit., p. 55.

My thanks to Juliana Engberg and Deirdre Summerbell for their insights.

ZENITH
TOSHIBA
Rembrandt

Video Review

THE WORLD AUTHORITY ON HOME ENTERTAINMENT

'LETHAL WEAPON 2' HITS HOME

MARCH 1990

ULTIMATE MEDIA ROOMS

A Top Designer's Secrets for Perfecting Your Home Theater

REVIEWS: 'Parenthood,' 'Indiana Jones,' 'Turner & Hooch'

LAB TESTS: Minolta & Ricoh Camcorders, Zenith TV

01271
$2.50/$2.75 CANADA

10th ANNIVERSARY SPECIAL

Going to the Movies, Staying at Home

By KENNETH LAWSON

IN the basement of a Victorian brownstone in the Park Slope section of Brooklyn is a playroom right out of the movies: a 1,000-square-foot Art Deco-style theater equipped with 10 red velvet seats and the latest technology.

Created by Theodore Kalomirakis, a film enthusiast and collector of about 4,000 cassettes and laser disks of films, the theater is reserved for entertaining friends. "Having my own theater is the only way I can indulge in the enjoyment of movies as they were meant to be enjoyed," said Mr. Kalomirakis, who emigrated from Greece 15 years ago to continue a career in film production.

When Mr. Kalomirakis bought the four-story 1880's brownstone in 1987 with two partners, Eric Seidman and John Thomsen, the three hired the architectural firm of Alvarado, Thrun & Maeda Associates to restore the house and plan a complete reconstruction of the underground space. The top two floors were converted to apartments; Mr. Kalomirakis lives on the ground and parlor floors.

"The idea was to restore the house as closely as possible to what it might have looked like a hundred years ago," said Mr. Kalomirakis, who is now an art director at American Heritage. Mr. Kalomirakis estimated that total cost of the improvements would come to $400,000, about a quarter of which went to the theater.

At the foot of the staircase to the basement, a red and blue neon sign announces "Roxy." Mr. Kalomirakis designed the sign as an updated version of the hand fan logo used by the legendary Roxy Theater in Manhattan on Seventh Avenue and 50th Street, which was razed in 1960.

An oblong gallery of film memorabilia is illuminated by soft lighting from Art Deco-style sconces. Programs from the original Roxy, left behind in the house by previous residents, are framed on the wall.

"I'd already decided to call my theater the Roxy," Mr. Kalomirakis said. "I just couldn't believe it when we found these old programs in the basement. It felt to me like the project was just meant to be."

Posters on the walls of the gallery celebrate long-forgotten B-movies like "The Farmer Takes a Wife," "Sis Hopkins" and "Happy Go Lovely." Toward the rear of the room, an old red candy machine dispenses chocolate bars for two nickels.

"Eventually, we'll have an entire concession area with a refurbished popcorn dispenser and a soda machine," Mr. Kalomirakis said. "In the rear, Dutch doors will lead to a service area where guests can have cocktails."

Archways mark the entrance to the auditorium of the theater. The custom-designed seats face a small, semicircular stage ringed by footlights. Behind it a curtain of red velvet conceals an eight-foot screen.

At the rear of the room are twin banks of high-tech equipment, including VHS and Beta recorders, a laser disk player, video processors that enhance color quality and picture sharpness, an audio equalizer, a surround-sound processor and amplifiers. Speakers are behind the screen and beside the seats.

Mr. Kalomirakis is disenchanted with the ways audiences experience films today. "You wind up in a multiplex, where you hear the sound of the movie playing next door," he said. "Your shoes get stuck in cola syrup. People talk behind you. The screens are tiny, and theater décor is nonexistent."

He added that although watching films at home may be more comfortable, it deprives viewers of what he feels is an important part of experiencing a movie. "The ritual of making the trip to the theater," Mr. Kalomirakis said, "seeing the marquee, waiting on line, buying popcorn and then watching the film straight through with no interruptions, is lost."

When movies became readily available on videocassette and laser disk, Mr. Kalomirakis began amassing a collection that now includes films in virtually all genres, old and new. His favorites are what he called "first-generation musicals," which were made from 1929 to 1932 and which preceeded the Busby Berkeley spectaculars.

Parts of Mr. Kalomirakis's collection highlight actors and directors. Currently, his tastes run toward Betty Grable, Deanna Durbin, Jack Oakie and Wheeler and Woolsey, the RKO comedy team of the 1930's. Other areas of his library focus on the works of the directors Alfred Hitchcock and Rainer Werner Fassbinder.

Mr. Kalomirakis is pleased by the reaction to his home theater from serious and casual movie fans alike.

"Sharing my films in this kind of environment is what makes this passion worthwhile," he said. "I cannot enjoy watching movies by myself; to me, they're meant to be seen with a crowd, in a collective event. It's no fun, though, if I'm preoccupied with being the host or the projectionist. That's just running a sideshow. I have to get in on the fantasy, too."

Smith-Miller + Hawkinson Architects. *Model Apartment*,
The Police Building New York, NY 1989.

SPACE

Publicity photograph for the Canon E80. This video camcorder features the new and unique "FlexiGrip," a combination grip and electronic viewfinder that rotates 180-degrees to make low- and high-angle shooting easy. By integrating grip and viewfinder into a single rotating unit, comfortable shooting is possible no matter what the angle.

How Television will Kill "Time"

New Worlds for Old

A Special Article written exclusively for "Television"

By SHAW DESMOND

Author of "Echo," "The Isle of Ghosts," "Ragnarok" (a story of the coming Armageddon), etc.

THREE of us sat in an upper chamber, remote from the fret and suck of the city's tides beneath. It was a queer room, full of curious drawings, of strange, almost chimerical, instruments, of screens—now milk white, now shining like mirrors under the summer sun.

From far below came the roar of the motor-buses; the cries of the newsboys; and the hoot of the taxis. All the *earth-bound* movement of our day.

And then, as we looked down, there came the ominous thrud-thrud of something above us, and a shadow as of a great black bird was flung upon our screen.

As the giant Handley-Page bomber roared overhead, with its attendant sprites, Fox scout-planes, turning and twisting in the central blue, one of the three, an engineer known in four continents, said quietly:

"All that below is passing. All that above is coming." And then, after a moment, the words, penetrating, low:

"*Look up! Not down!*"

What the motor car was to the horse, and the aeroplane to the motor-car, the "televisor" will be to the aeroplane . . . and to the world. It is only because our imaginations are *earth-bound* that we have not seen it.

For it is television that is going to wipe out space and, above all, wipe out *time*.

In the world of the future, time, as such, will have no existence. It will be as dead as Adelphi melodrama or crinolines. We, human beings, will no longer reckon our little lives by watches and clocks, but by *events*.

Before twice a dozen of years have petered out into eternity we may have neither motor-bus nor taxi in our streets. As for the newsboy—his yell will by then be extinct as the yell of his forbear, the Red Indian.

For, as will be seen, the newspaper will be literally televised into the home. Simultaneous publication of the world's great news-sheets will be such a common-place, through the medium of the "televisor," that the tutored African savage, an he will, will be able to read the London *Times* in the streets of Johannesburg or Durban *at the moment of its printing* in the present capital of the British Empire and of the world. "Present" I say—for the world's capital by that time may be New York or Toronto or Timbuctoo! Things change so quickly to-day.

For the world is moving. As old Galileo said: "It moves."

Nay, more, the world as we knew it, and know it, is passing never to return. The "televisor" is going so to obliterate the old landmarks that were we to come back within, say, fifty years we should not know the world we had left, save by its buildings—and not by many of those, for the skyscraper is coming.

* * *

I walked out from that meeting in that room, where the history of our day is being made, to London's new circular underground station at Piccadilly.

In front of me was a big lighted screen. Over the top:

WHAT'S THE TIME?

It was prophetic.

There, within ten minutes of my friend saying "Look up!" and my own prophecy of the abolition of time by seeing at a distance, I found myself looking at a screen before

"Three of us sat in an upper chamber . . . full of curious instruments, of screens—now milk white, now shining like mirrors under the summer sun."

I Remember Television. . .

David Tafler

When writers from the not too distant past looked into the future, they somehow accurately envisioned the future of a time that has yet to come. Windows that they opened revealed Tom Swift's moving image projector, Dick Tracy's televideo watch, and H.G. Wells' time machine. These time voyagers discovered a narrow, transient breach somewhere between a future reality and an author's imagination. Their fiction succeeded in dramatically tracking the tentative thread that connects the past with the not so immediate future.

In the world of non-fiction, meanwhile, engineers, technicians, and other media utopians, seduced by the dramatic possibilities of video transmission during the late 1920s and early 1930s, made another sort of leapfrog projection, by describing the electronic environment of the late 1990s or early twenty-first century in their writings. They did not cover, however, the forty interim years during which television proliferated and evolved from a novelty to an everyday reality. During that interim period, television brought technological advances, but it resembled an old paradigm. Despite its visual enhancement, television maintained the audio status quo of radio. Though radio changed and the movies eventually became a part of the tele-equation, the old radio paradigm prevailed.

With television now the prevailing force, will this sort of developmental pattern repeat itself in future prognostications? Insofar as the television set appears in every conceivable environment and television events saturate everyday experience, future speculations on the media will obviously evolve from contemporary television paradigms. Undoubtedly, the television prototype of the present will model the immediate future. The future of television, however, will transcend "television."

For the time voyager, looking to a far millennia makes fewer demands and, therefore, is a more inviting task than predicting the subtle and misleading trends of the next one hundred years. Imperceptible shifts often obscure dramatic changes. Moreover, recognition of new paradigms demands the evolution of new systems of thinking. While the visionary looks outward to observe the waves of change embracing the world, a considerably more occluded view blocks the multidimensional perspective required for prognostication on the inside. Not surprisingly, however, this interior space also harbors the dynamic temporal forces of greatest impact on those living today and their immediate descendants. Those forces, in turn, shape the extended future. In short, the non-utopian constraints of the home anchor the imaginary horizon.

More than a passive pastime, television stitches together the many threads of today's dynamic social-economic-political environment. More importantly, as an affective system which satisfies deep seated human needs, television articulates human desire and drives society's material appetite. It fuels the imagination, drives the economy, and deflates guilt. In a society which simultaneously heralds and deflates notions of individuality as well as community, the electronic highway captures and reflects those contradictions and channels its constituents' frustration and anxiety. The basic tenets of institutional and individual competition motivate television's existence and determine its metamorphosis. The prophetic vision must fathom these conditions and comprehend how television weaves together the benign and explosive forces operating within the social body.

Within the home, windows to the future often pass across the screen. Many years ago, in the interim period before public broadcasting stations occupied channel thirteen in New York City, this station broadcast a somewhat bizarre truncated schedule and repertory. During the non-programmed daytime hours, a fixed

YOU'LL BE AN ARMCHAIR COLUMBUS!

You'll sail with television through vanishing horizons into exciting new worlds. You'll be an intimate of the great and near-great. You'll sit at speakers' tables at historic functions, down front at every sporting event, at all top-flight entertainment. News flashes will bring you eye-coverage of parades, fires and floods; of everything odd, unusual and wonderful, and bring them to you just as though you were on the spot. And farsighted industry will show you previews of new products, new delights ahead.

All this—the world actually served to you on a silver screen—will be most enjoyably yours when you possess a DuMont Television-Radio Receiver. It was DuMont who gave really *clear* picture reception to television. It will be DuMont to whom you will turn in peacetime for the finest television receiving sets and the truest television reception...the touchstone that will make you an armchair Columbus on ten-thousand-and-one thrilling voyages of discovery!

DUMONT *Precision Electronics and Television*

ALLEN B. DuMONT LABORATORIES, INC., GENERAL OFFICES AND PLANT, 2 MAIN AVENUE, PASSAIC, N. J. • TELEVISION STUDIOS AND STATION W2XWV, 515 MADISON AVENUE, NEW YORK 22, N. Y.

image would frequently linger on the screen while a local radio station provided the sound. Ironically suggestive of Nam June Paik's work, the viewer often saw fish swimming across a screen-size fish tank. At night, movies took over the screen.

One night, the evocative image of KRONOS, from the 1957 science-fiction film of the same name, replaced the fish. Along an unspecified "time capsule" typified by a primitive stretch of Mexican coast, a North American traveller goes to shave, casually looks out the bathroom window of his simple shack, and unexpectedly beholds "Kronos," an outrider of the future. Through the breach, he observes the natives hysterically worshipping an enormous metallic, moving object which would later resemble one of the World Trade Center towers in New York.

Kronos, a technological monster, represents the trauma of an intractable transition. Its namesake Cronus, youngest of the Greek Titans, deposed his father Uranus, married his sister Rhea, and later himself succumbed to Zeus. The film's Kronos, a robot colossus capable of absorbing the Earth's energy, symbolizes the revolving door of good and bad technology and its penetration into the "home." While the antenna atop the World Trade Center tower beams the transmitted television signal to the New York metropolitan area, television, the window in the protagonist's shack, merely represents the transducer or conduit feeding the impulses of this force to its satellite constituents. For the time being, it remains a controllable pathway. The future of the tele-apparatus in the home, however, suggests the erosion of that control.

Despite the voracity of change, technological and physiological limits restrain evolutionary innovation. Nothing alters dramatically. Systems reorganize within economically determined margins. Human desire and imagination govern the interface between the technological and physiological body. Outside the cybernetic imagination, the body remains the same.

On a commuter train heading for New York, a three year old girl tells her mother, "when I get home, I want to lie down on the couch, watch TV, and have something to eat." For this little girl and many other people, television persists as the emotional umbilical cord, the sentient or conscious outlet for hallucinatory activity, the ameliorative for social isolation, the purlieu for loss of sleep. It anchors fixed patterns of daytime and nighttime activity, and channels residual emotional excess. Despite advances in technology, the human condition remains the same; human needs endure despite changing external exigencies.

In time, however, the paradigms governing the human condition change. As the human body accommodates to change, the tele-apparatus increasingly reflects the shift from an industrial to an information society, from the muscles of the industrial body to the nerves of the corporate state. With the shift, the parameters for understanding evolutionary change become meaningless. A search begins for ways of defining the new systems and subsystems, the filters, catalysts, and synaptic connections of a new social foundation—for ways of measuring true and meaningful change.

Present thinking still prescribes the old feudal borders between technological applications. Economic and social functions denote the differences between the public and private sphere, the workplace and the home. Unfortunately, looking into the future from a dated platform produces at best a crippled perspective. At worst, it describes a retrograde environment. In order to avoid the fallacy of charting a utopian odyssey, forecasts of the future must begin with the development of a new vocabulary which evolves from the dramatic changes that transpire over time.

The old vocabulary transforms alongside the erosion

Publicity photograph for Nintendo "Power Pad" video game, 1988.

of traditional boundaries separating spheres of economic and social activity. On the horizon, day-to-day functions no longer distill into separate economic jurisdictions. Once these functions overlap, conditions in the home will necessarily change. The home becomes cyberspace, a cybernetic station integrating work, home work, and pleasure—a place for friends, colleagues, associates, and classmates, not to mention family. As part of the conversion, experiences within the home expand through the now programmable, computer driven environment. Under the new rubric of the homestation, television quickly fades away from its former model as a late twentieth century artifact. Becoming a dated concept, television follows on the heels of the phonograph to the lingual oblivion of canned nostalgia, merely suggestive of its pioneer past.

In its place, only temporarily, there resides a revitalized cube, a monitor, a prolongation of a dated commodity. Initially, the multi-functional monitor replaces the television set as communication systems combine through a field of greater synchronization. Computer terminals merge with video screens; CD Roms link into database networks and VCRs. Applications overlap. The camcorder, telephone, computer, cable/direct satellite/on-line monitor interface in the living room, work room, or kitchen studio.

Hierarchical differences, nevertheless, still exist in the home. Each home equips its telecommunication stations differently. Buffers exist that limit operations of artificial intelligence to those experienced and educated spect-actors equipped to enter the appropriate pathways. Matching processing systems with their biological counterparts, power governors or controlling devices process input and limit control to suitable agents.

In the new home, input devices proliferate. Those devices determine the configuration of the room(s), walls, windows, artificial light, and heat sources. Each device has its own FAX or modem address. As alternative entranceways to the world, they compete with the physical doors and hallways as conduits for arrival and departure. Already, telephone usage forecasts these developments.

Eventually, the box explodes. First, the surface flattens out squaring off the edges and corners of the no longer bulging television tube. So far so good. But then the screen disappears. The screen becomes an abstraction, a module, tube, or tunnel. All of a sudden, the information on the screen escapes to the light of day. In its place lies a gaping hole, an opening for a new discourse, waiting, beckoning, inviting, simultaneously threatening and rejuvenating. Once more through the looking glass, every man or woman enters wonderland. The body becomes inseparable from the box. The spectator becomes interactor—a helmeted walk(wo)man—plugging into the walls and donning his or her data glove.

In this territory of virtual conception, passageways substitute for the screen and become the vital stitch in an electronic weave. Leaning back, the spect-actor projects onto the mirror a transparent interface between his or her interior and multidimensional exterior. Charting his or her way through the information arena, the spect-actor integrates certain recognizable features of the grid, at the same time consciously erasing what lies beyond the immediate domain of his or her utilization and/or fascination. Following the first directive, the spect-actor maps out his or her position. From that junction, a particular pathway usually beckons. Otherwise, a number of recursions or formulaic schemata present substitute possibilities. Surprisingly, resolution may come through a loop procedure. Alternately, an endless maze of limitless corridors may eventually exhaust the explorer. The explorer then moves off the track and temporarily parks in a way station to recharge the spirit or revitalize the terms of the search.

In an evolutionary leap, a metaphysical shift has moved the center away from the screen. The electric armchair, not the screen, anchors the new interface. The emphasis rests on the chair which embraces the console. Beyond mere cognition-perception, the point of interaction with the screen lies at the spect-actor's fingertips. To the undiscerning, this new paradigm bears a remarkable resemblance to the twentieth century archetypal figure situated at the controls of his or her automobile, directing the vehicle and perpetually reconstructing an array of auditory-visual events on the windshield, radio, tapedeck, CD player and CB. Other references emerge from fictional prototypes. As in H.G.Wells' *The Time Machine,* the ages of now past and projected future civilization peel away from the ephemeral bubble which encloses the wide-eyed time traveller's floating cushioned console and chair. More recently, the model of the corporeal platform presents itself on *Star Trek,* in both the original and *The Next Generation,* where the focal center rests on the central chair(s) occupied by the Captain and his commander(s).

Nevertheless, key differences in the underlying experience dictate the evolutionary leap to the electronic chair of the future. The telectronic chair escapes the mechanical restraints and regulatory laws of the automobile, time machine, and starship. Responsibilities attendant to moving a vehicle fall away when nothing but the mind triggers the individual's flight.

Through the wireless remote's projected infra-red rays, the spect-actor's perceptions complete the interactive cinematic loop. Advancing forward, the technical extension shapes the next phase of human evolution toward the development of a super organic creature that exceeds the boundaries of the flesh and circuit board. These altered states merge the perceptual-cognitive operation of two integrated nervous systems. The cybernetic android combines the machine interface with a human form.

Somewhere along the way, questions arise as to the other spectators in the room. Will they occupy another module beyond the console, somewhat comparable to the late twentieth century star voyeur cruising on the other side of the screen? Will a seamless experience of desire and expectation separate the active grazer from the outside scavenger who picks up the reusable discards compiled from the other's actions? No longer equal, each potential interactor will either compete for control or separate into their own modular units. The living room disappears. The move from windows to passageways transforms the house into a multi-dimensional earth station.

Historically, does this mean the end of the family home? With the demise of the nuclear homestead as a central organizing force, there comes a resurgence of the community, a return from suburbia. The body resists for the sake of love and procreation, but the tether unravels in myriad ways. Linked through networks, the new community does not resemble a return to the old nineteenth century paradigm. In its place, the simultaneity of networked experience shared by users around the globe becomes a new tabernacle for the virtual community.

Two irreconcilable movements form the axes of the new world. One, a movement toward greater synchronization works to evolve network strategies for segmenting and/or fragmenting communication. Institutions emerge modeled on corporate formulas for presenting, storing, and transmitting information. The other movement mediates the fracturing of the community by recognizing the terminus of each signal. At the edge of the screen, the absent figure represents an outland. The screen, in turn, functions as a barrier reef. Since the protection of the system guarantees the fracturing of the community, the second movement operates as a third voice generating meaning as well as light. Since the body remains human, each movement balances the other, accommodating inevitable biological and psychological changes in the human species.

It does not seem likely that our species will arrive at a utopian juncture where society can successfully accommodate the electronic fragmentation of behavior and responsibility. Nevertheless, it remains a fundamental necessity for survival. Already forecast, the future merely involves modifying the life activity on board an imaginary star ship to the limitations and constraints of an entire planet. The alternative requires violating trends that have become endemic. Those possibilities also exist. ■

An early cathode ray tube, c. 1939.

The End of the Television Receiver

Margaret Morse

The box on display in the living room—the body of television—has been technologically superseded by video projectors and liquid crystal displays. Television no longer requires a three-dimensional body of its own. Though TV sets—be they mahogany or plastic—proliferate numerically into every room in the house and qualitatively beyond, into schoolrooms, courtrooms, legislatures, churches and stadia, the television receiver is already a nostalgic object. The fantasy of collective individualism embodied in the suburb-television-freeway-mall complex[1] may be displaced as well by an immaterial culture projected in light, commonly known as virtual reality. The following speculates on the imbrication of disembodied television with the computer and the cultural implications of cyberspace, or computer-generated virtual worlds as metaphors which we can walk through, touch, hear and see.[2]

It is true that we ordinarily think of the TV set less as a body than as container or frame for transmitted, not source, images and sounds. Indeed it is *as if* (but not really) the set were hollow within like a microwave or a box of salesman's samples or a stove for cool fire. Inside the hollow television, the ultimate box, is a personal reliquary for fetish objects or sacra at the crossroads of everyday life, the commodity world and our common culture.

But the TV box is also a shape the size of an ersatz person in a medium shot or close-up. Like minimal sculpture, the box confronts the visitor in a shared space and real time of changing light and shadow[3]—except the TV sculpture is its own light source and manufactures its own temporal events. And, in the case of television, the sculpture does more than simulate human shape and scale, it is a container for talking heads: a machine-subject importunes us with the power of speech. The spot in front of the television is the virtual space of a simulated conversation we share with our subject-machine, a space which we can occupy or not, at any time. The TV set as utility for story-telling slaves supplies charms against loneliness on demand, offering fictions of a world (real or imaginary) shared without responsibility, at the price of imbricating our dreams with the fetishism of commodities.[4]

The television receiver as box is gradually being displaced by the computer display terminal, integrated with the telephone, the typewriter and drawing tablet, the library, the game and bulletin board. And, even now, you can install a board and program to receive television on the computer screen. Here the ersatz person has set up shop again, interacting with us with ever more sophistication, but nonetheless virtually. The body of the computer becomes a projection of our second self, co-mingled with the projections of other subjects in messages from an ambiguously inner-outer other world. Video projection allows this other world and mixed self-other subjectivity to be projected out of the box onto our real conditions of existence.

The logic of the box as container is quite different from the logic of projection. Projected or liquid crystalline images are surfaces of light, not bodies. Lovely fine-grained phantoms in saturated colors lack a body or frame to contain them or a luminous path to betray their origin. Their scale can be superhuman in larger-than-life projection, while liquid crystals can be miniatures, worn like jewelry. Like shadows of stained glass without the need of window or sun, they are easily wrapped over any architectural feature or object. An architecture of giant or miniature skins of light can be condensed over walls or objects in real space: even now, video walls can move and speak with human voices in the mall. Images need not coincide with their frame or support: with the help of a computer, an image can be broken over any shape in any size, whole or in a fragment.

But skins of light don't necessarily require any reality support: On the freeway, we can soon anticipate the appearance of the virtual video screen or "head up display" which will float in a driver's field of vision like a freeway sign.[5] Virtual television can float in our space like a squeeze frame over the anchor's shoulder in the news. It is *as if* we were virtually *inside* the presentation of television events. Our imaginary move inside the television has been anticipated daily for decades by all the *2001*-style "star gate" sequences from logos and Max Headroom to *Peewee's Playhouse* that seem to zoom us into the television set. Star gates have already migrated into real space in video art: floors may open like rabbit holes into animate worlds—at least virtually.[6]

Phantoms or skins of light do not inhabit real time. Manipulated by camcorder and VCR, images can be repeated, sped up, slowed down, edited (in real or unreal time) and submitted to substitution, multiplication or undeath (i.e. reversible unbeing) with a single remote gesture. Furthermore, sound is freed from an acoustic perspective and source, to simulate itself in other realms in sampled and layered complexity. As a result, the recorded world may be retrievable and instantly recognizable, displaceable, stackable in thin wafers of dense sound, delightful. So, a fantasy of a world without gravity, space or time which we inhabit *bodily* is elaborated by the disappearance of the box.

It is *as if* our symbolic realm were no longer contained in and by objects, the written page, or echoed in acoustic space. Our box of symbols and words is emptied out, spilling husks of speech and gestures out into the air. Inside has become outside. And outside has become inside, without a frame to call us home from dream time.[7]

We may be inside television, but we are not contained: this inside is a world without edges or ends, a space without place, where planes overlap and intersect without boundaries or frames. The prospect looms of psychic regression to a stage even earlier than the cave. To inhabit metaphors made of light may put the most fundamental distinctions—present/absent, living/dead—at risk in a world of mundane magical transformations and commodity terror. Without the shelter of the niche or box *from* as well as for words, language can become undead, images can become vampires.

But though it seems that projected worlds are psychically preparing us for the weightlessness of outer space—our bodies act nonetheless in real space and irreversible time on earth. It may be fair to say that there is an unacknowledged and adverse relation between the image-world and the earth it displaces, now ugly and deprived of desire. This earth enmity is not assuaged by make-overs with image skins advertising for the environment. Coming to terms with virtual reality is a new task of art. Perhaps only the contrast of kinetic and virtual information, of gravity and phantom light can locate a hard place on which to stand—a task I foresee for video installation art.[8] ■

NOTES

1. See my "Ontology of Everyday Distraction: the Freeway, the Mall and Television," in: *Logics of Television: Essays in Cultural Criticism*, ed. Patricia Mellencamp, University of Indiana Press, forthcoming.
2. Computers are the source of virtual realities in the strictest sense. Cf. Timothy Binkley,"Camera Fantasia: Computer Visions of Virtual Realities," *Millenium 20/21* (Fall/Winter 1988-1989), 6-45. It should be added that the virtual camera which produces a projected world is not compelled to sim-

ulate the properties of perspectival space and linear time. Yet, computers require visual metaphors (i.e. television and beyond, to a multidimensional cyberspace) to make formless data perceivable and comprehensible. See Peter H. Lewis, "Put on Your Data Glove and Goggles and Step Inside," *New York Times,* May 20,1990, p. F8. Currently available software (Autodesk, Inc.) enables users to "walk through structures which exist only as data within the computer." Of course, an interface is required, an inversion of Siegfried's helmet for making the wearer invisible—goggles which make the world invisible and displace the visual field with a computer-generated vision. Data gloves allow one to feel information as a metaphor given a literal shape, if only a virtual existence. The article anticipates virtual workplaces organized, sad to say, by "the office metaphor."

3. See Michael Fried in "Art and Objecthood," *Aesthetics Today,* eds. Morris Philipson and Paul Gudel (New York: NAL, 1980), pp. 214-239, for a seminal description of the anthropomorphic and theatrical qualities of minimalist sculpture.

4. See my description of "the fiction of discourse" in: "Talk, Talk, Talk: The Space of Discourse in Television News, Sportcasts, Talk Shows and Advertising," *Screen* 26 No. 2 (March-April 1985), 1-15.

5. Edward E. Duensing, "Television on the Move: In-Car Video Screen Small But Critics Question Safety," Los *Angeles Times,* September 11, 1989, Part II, p. 3. The article describes the new technology recently patented by Jay Schiffman of Auto Vision Associates in Ferndale, Michigan.

6. See, for instance, the installation work of Judith Barry, *Maelstrom: Max Laughs and Adam's Wish,* both 1988/89, with floor and ceiling star gates, respectively.

7. See the discussion of Joseph Cornell's boxes in Edmund Burke Feldman's chapter on "Niches, Boxes and Grottoes," in *Varieties of Visual Experience* (New York: Prentice-Hall, 1987), p. 347, where "the frame takes us home." Feldman also refers to Kenny Scharf's *Extravaganza Televisione* 1984 as "the ultimate box," p. 349.

8. For further speculation on the cultural function of video installation art, see my article "Video Installation Art: the Body, the Image and the Space-in-Between," in: *Illuminating Video,* eds. Doug Hall and Sally Fifer, Aperture Press, forthcoming, November 1990.

Notes on the Contributors

Serafina K. Bathrick teaches in the Department of Communications at Hunter College in New York City.

Andrew Behar and *Sarah Sackner* are filmmakers and the principals of Behar-Sackner Communications in Sherman Oaks, California.

William L. Bird is the curator of the exhibition "American Television from the Fair to the Family 1939-1989" at the National Museum of American History, Smithsonian Institution.

Ed Bowes is a writer, producer and director of television programs. His most recent show is *Spitting Glass.* He is currently developing a sixteen part series which considers everyday life.

Les Brown is the Editor-In-Chief of *Television Business International* and the author of several books on television.

John G. Hanhardt is Curator, Film and Video at the Whitney Museum of American Art.

Maud Lavin is a writer and cultural historian. Her essays on design have appeared in *Art in America, Artforum, Elle,* and the Walker Art Center catalogue, *Graphic Design in America.*

Ehrick V. Long teaches at Hunter College in New York City and is working on his Doctorate in Musicology.

Joshua Meyrowitz is a Professor at the University of New Hampshire and the author of *No Sense of Place: The Impact of Electronic Media on Social Behavior* (Oxford University Press, 1985).

Margaret Morse is an Assistant Professor of Critical Studies at USC in Los Angeles, California. She writes on television, video, installation art and mass culture. Her book on "Television Reality" is forthcoming from Indiana University Press.

Jane Root, a TV researcher and journalist in England, is the author of *Pictures of Women: Sexuality* and *Open the Box.*

Lynn Spigel teaches in the Department of Communication Arts at the University of Wisconsin in Madison, Wisconsin.

David Tafler, a member of the faculty of the University of the Arts in Philadelphia, Pennsylvania, has written extensively on expanded interactive media, independent film, video, television and the arts.

Alice H. Yang writes on issues of contemporary visual art. She is Curatorial Coordinator at The New Museum of Contemporary Art, New York.

Staff

Kimball Augustus
Security
Richard Barr
Volunteer Coordinator
Virginia Bowen
Preparator/Assistant to the Registrar
Jeanne Breitbart
Assistant, Curatorial Department
Susan Cahan
Curator of Education
Helen Carr
Limited Editions Coordinator
Luis De Jesus
Curatorial Intern
Antonette DeVito
Director of Planning and Development
Sarah Edwards
Membership Coordinator
Russell Ferguson
Librarian/Special Projects Editor
Angelika Festa
Assistant to the Librarian
Phyllis Gilbert
Docent/Museum Tour Coordinator
Janet Gillespie
Bookkeeper
Roberta Green
Trips Coordinator
Maren Hensler
Collectors' Programs
Ellen Holtzman
Managing Director
Elon Joseph
Security
Patricia Kirshner
Operations Manager
Zoya Kocur
High School Educator
Sowon Kwon
Library/Special Projects
Alice Melendez
Receptionist
Clare Micuda
Assistant to the Director
France Morin
Senior Curator
Marisol Ng
Membership
Barbara Niblock
Administrator
Sara Palmer
Director of Public Affairs
Debra Priestly
Registrar
Howard Robinson
Security
Wayne Rottman
Gallery/AV Coordinator
Aleya Saad
Special Events Coordinator
Gary Sangster
Curator
Margaret Seiler
Publications Coodinator
Abigail Smith
Assistant Administrator
Susan Stein
Admissions/Bookstore Coordinator
Laura Trippi
Curator
Marcia Tucker
Director
Alida Vega
Public Affairs Intern
Alice Yang
Curatorial Coordinator
Lisa Zywicki
Curatorial Administrative Assistant